Advanc

With Applications in Physics,
Chemistry, and Beyond

ADVANCED CALCULUS EXPLORED

WITH APPLICATIONS

IN

PHYSICS, CHEMISTRY, AND BEYOND

Hamza E. Alsamraee

This page is intentionally left blank.

Copyright © 2019 by Hamza E. Alsamraee. All rights reserved. No part of this book may be reproduced or distributed in any form, stored in any data base or retrieval system, or transmitted in any form by any means—electronic, mechanical, photocopy, recording, or otherwise—without prior written permission of the publisher, except as provided by United States of America copyright law. For permission requests, e-mail the publisher at curiousmath.publications@gmail.com.

First edition published November 2019

Book cover design by Ayan Rasulova

Typeset using LaTeX

Printed on acid-free paper

ISBN 978-0-578-61682-7

www.instagram.com/daily_math_/

To my parents.

To numerically confirm the many evaluations and results throughout this book, various integration and summation commands available in software produced by Wolfram Research, Inc. were utilized. Moreover, virtually all illustrations and graphs were produced by software produced by Wolfram Research, unless specified otherwise. Specifically, Wolfram Desktop Version 12.0.0.0 running on a Windows 10 PC. As of the time of the release of the book, this is the latest release of Wolfram Desktop. The commands in this book are standard and are likely to continue to work for subsequent versions. Wolfram Research does not warrant the accuracy of the results in this book. This book's use of Wolfram Research software does not constitute an endorsement or sponsorship by Wolfram Research, Inc. of a particular pedagogical approach or particular use of the Wolfram Research software.

Contents

About the Author   15

Preface   23

I   Introductory Chapters   **27**

1   Differential Calculus   **29**
    1.1   The Limit 30
        1.1.1   L'Hopital's Rule 36
        1.1.2   More Advanced Limits 39
    1.2   The Derivative 45
        1.2.1   Product Rule 46
        1.2.2   Quotient Rule 48
        1.2.3   Chain Rule 49
    1.3   Exercise Problems 56

2 Basic Integration   59

   2.1 Riemann Integral 60

   2.2 Lebesgue Integral 63

   2.3 The u-substitution 65

   2.4 Other Problems 81

   2.5 Exercise Problems 98

3 Feynman's Trick   101

   3.1 Introduction 102

   3.2 Direct Approach 103

   3.3 Indirect Approach 128

   3.4 Exercise Problems 131

4 Sums of Simple Series   135

   4.1 Introduction 136

   4.2 Arithmetic and Geometric Series 136

   4.3 Arithmetic-Geometric Series 141

   4.4 Summation by Parts 146

   4.5 Telescoping Series 152

   4.6 Trigonometric Series 159

   4.7 Exercise Problems 163

CONTENTS

II   Series and Calculus · · · 165

5   Prerequisites · · · 167

- 5.1   Introduction · · · 168
- 5.2   Ways to Prove Convergence · · · 173
  - 5.2.1   The Comparison Test · · · 173
  - 5.2.2   The Ratio Test · · · 173
  - 5.2.3   The Integral Test · · · 177
  - 5.2.4   The Root Test · · · 181
  - 5.2.5   Dirichlet's Test · · · 184
- 5.3   Interchanging Summation and Integration · · · 185

6   Evaluating Series · · · 191

- 6.1   Introduction · · · 192
- 6.2   Some Problems · · · 193
  - 6.2.1   Harmonic Numbers · · · 204
- 6.3   Exercise Problems · · · 213

7   Series and Integrals · · · 215

- 7.1   Introduction · · · 216
- 7.2   Some Problems · · · 216
- 7.3   Exercise Problems · · · 230

8 Fractional Part Integrals — 233

- 8.1 Introduction 234
- 8.2 Some Problems 235
- 8.3 Open Problems 255
- 8.4 Exercise Problems 255

III A Study in the Special Functions — 257

9 Gamma Function — 261

- 9.1 Definition 262
- 9.2 Special Values 262
- 9.3 Properties and Representations 263
- 9.4 Some Problems 270
- 9.5 Exercise Problems 275

10 Polygamma Functions — 277

- 10.1 Definition 278
- 10.2 Special Values 279
- 10.3 Properties and Representations 280
- 10.4 Some Problems 282
- 10.5 Exercise Problems 294

11 Beta Function — 297

- 11.1 Definition 298
- 11.2 Special Values 299
- 11.3 Properties and Representations 299
- 11.4 Some Problems 304
- 11.5 Exercise Problems 311

12 Zeta Function — 313

- 12.1 Definition 314
- 12.2 Special Values 314
- 12.3 Properties and Representations 319
- 12.4 Some Problems 328
- 12.5 Exercise Problems 337

IV Applications in the Mathematical Sciences and Beyond — 341

13 The Big Picture — 343

- 13.1 Introduction 344
- 13.2 Goal of the Part 345

14 Classical Mechanics — 347

- 14.1 Introduction 348

14.1.1 The Lagrange Equations 348

14.2 The Falling Chain 349

14.3 The Pendulum 355

14.4 Point Mass in a Force Field 359

15 Physical Chemistry   365

15.1 Introduction 366

15.2 Sodium Chloride's Madelung Constant 372

15.3 The Riemann Series Theorem in Action 374

15.4 Pharmaceutical Connections 379

15.5 The Debye Model 380

16 Statistical Mechanics   385

16.1 Introduction 386

16.2 Equations of State 387

16.3 Virial Expansion 389

16.3.1 Lennard-Jones Potential 390

16.4 Blackbody Radiation 392

16.5 Fermi-Dirac (F-D) Statistics 397

17 Miscellaneous   405

17.1 Volume of a Hypersphere of Dimension N 406

17.1.1 Spherical Coordinates 406

17.1.2 Calculation 408

17.1.3 Discussion 411

17.1.4 Applications 413

17.1.5 Mathematical Connections 414

V Appendices   417

Appendix A   419

Appendix B   425

Acknowledgements   429

Answers   431

Integral Table   439

Trigonometric Identities   443

Alphabetical Index   447

About the Author

Hi! My name is Hamza Alsamraee, and I am a senior (12th grade) at Centreville High School, Virginia. I have always had an affinity for mathematics, and from a very young age was motivated to pursue my curiosity. When I entered a new school in 7th grade after moving, I encountered some new mathematics I was unequipped for. Namely, I did not know what a linear equation even was! I was rather low-spirited, as I was stuck in an ever-lasting loop of confusion in class.

My mother and father soon began teaching me to the best of their ability. Fortunately, their efforts were effective, and I got a B on my first linear equations test! It was a huge improvement from being totally lost, but I wanted to *know* more. I did not care much about the grade, but I did care that I did not completely master the material.

I soon entered in a period of rapid learning, delving into curriculums significantly beyond my coursework simply for the sake of mastering higher mathematics. It seemed to me that the more I explored the field, the more beautiful the results were.

I quickly got bored of the regular algebra and geometry problems, and wanted to know if there was more to mathematics. I browsed the web for "hard math problems," and stumbled across this integral:

$$\int_{-1}^{1} \sqrt{\frac{1+x}{1-x}}\, \mathrm{d}x = \pi \tag{1}$$

Well, I knew what π is, but what is that long S symbol on the left side? So, I browsed the web again, but this time asking what that "long S symbol" is. It is an integral! Cool, I thought, but what does that even mean?

Soon started a period where I was almost obsessed with those mathematical creatures! The evaluation of integrals and series quickly became a hobby, leading me to challenge myself everyday with a new integral or series that "looked" like it might yield a nice closed form.

As I delved deeper into the subject matter, I discovered the well-known special functions such as the gamma and zeta functions. Being a physics enthusiast, I was fascinated by their applications in physics as well as in other scientific disciplines. I found my love for mathematics and science converge, and was determined to cultivate this passion.

Ever since then, I began collecting results and solutions to various problems in the evaluation of integrals and series. It was only about a year ago, due to a suggestion of one of my close friends, that I thought about compiling my results into a cohesive curriculum. I remembered my early days in doing these sorts of problems, and my frustration at the lack of quality resources. It was then that I became determined to write this book!

After I began writing the book, I began wondering how many individuals were genuinely *interested* in this type of mathematics. As an experiment, I set up a mathematics Instagram account by the name of *daily_ math_* to test the reaction to some of the book's problems. After an overwhelmingly positive response, I was more motivated than ever to finish this book! At the time of the publishing of this book, the account has garnered over 40,000 followers globally, from middle school students to mathematics PhD's.

Proof of (1)

This integral is truly special to me, as it marked the beginning of a period of immense investment into my passion, mathematics. Thus, I have compiled a list of proofs that I derived over the years. Even though all these proofs were arrived at independently, they exist in the literature since this integral is rather common.

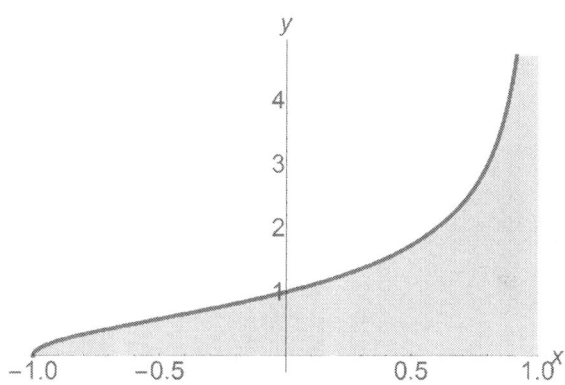

Figure 1: Graph of the integrand of (1)

Proof 1. We will begin by splitting this integral from -1 to 0 and from 0 to 1 to get:

$$I = \int_{-1}^{1} \sqrt{\frac{1+x}{1-x}} \, dx$$

$$= \int_{-1}^{0} \sqrt{\frac{1+x}{1-x}} \, dx + \int_{0}^{1} \sqrt{\frac{1+x}{1-x}} \, dx$$

Using the substitution $x \to -x$ on the first integral,

$$I = -\int_{1}^{0} \sqrt{\frac{1-x}{1+x}} \, dx + \int_{0}^{1} \sqrt{\frac{1+x}{1-x}} \, dx$$

$$= \int_{0}^{1} \left(\sqrt{\frac{1-x}{1+x}} + \sqrt{\frac{1+x}{1-x}} \right) dx$$

$$= \int_{0}^{1} \frac{(1-x) + (1+x)}{\sqrt{(1-x)(1+x)}} \, dx$$

$$= 2 \int_{0}^{1} \frac{1}{\sqrt{1-x^2}} \, dx \qquad (2)$$

Since

$$\frac{d \arcsin x}{dx} = \frac{1}{\sqrt{1-x^2}}$$

We have:

$$I = 2[\arcsin x]_{0}^{1}$$

$$= \boxed{\pi}$$

An alternate way to compute (2) is by the substitution $x = \sin u$, $dx = \cos u \, du$:

$$\int_{\arcsin 0}^{\arcsin 1} \frac{\cos u}{\sqrt{1 - \sin^2 u}} \, du$$

$$= \int_{\arcsin 0}^{\arcsin 1} du$$

$$= \arcsin 1 - \arcsin 0 = \frac{\pi}{2}$$

∎

Proof 2. Consider multiplying (1) by $\dfrac{\sqrt{1+x}}{\sqrt{1+x}}$

$$I = \int_{-1}^{1} \frac{1+x}{\sqrt{1-x^2}} \, dx$$

$$= \int_{-1}^{1} \frac{1}{\sqrt{1-x^2}} \, dx + \int_{-1}^{1} \frac{x}{\sqrt{1-x^2}} \, dx$$

For the second integral, we let $u = 1 - x^2$, $-\frac{du}{2} = x \, dx$ to get:

$$I = \pi - \frac{1}{2} \underbrace{\int_{0}^{0} \frac{1}{\sqrt{u}} \, du}_{=0}$$

Which can be easily argued from the fact that the integrand in the second integral is an odd function on the interval $(-1, 1)$.

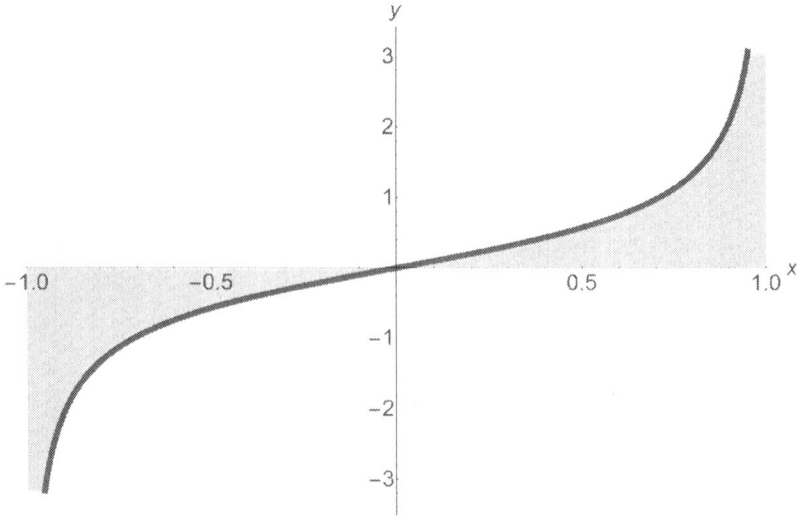

Figure 2: Graph of $y = \dfrac{x}{\sqrt{1-x^2}}$ on $(-1, 1)$

We therefore obtain:

$$I = \pi$$

Proof 3. Consider the integral reflection property,

$$\int_a^b f(x) \, \mathrm{d}x = \int_a^b f(a+b-x) \, \mathrm{d}x$$

Applying that to (1) gives:

$$I = \int_{-1}^{1} \sqrt{\dfrac{1-x}{1+x}} \, \mathrm{d}x \qquad (3)$$

Adding (3) and (1) gives:

$$2I = \int_{-1}^{1}\left(\sqrt{\frac{1-x}{1+x}} + \sqrt{\frac{1+x}{1-x}}\right)\,\mathrm{d}x$$

$$= \int_{-1}^{1}\frac{2}{\sqrt{1-x^2}}\,\mathrm{d}x$$

$$= 2(\arcsin(1) - \arcsin(-1)) = 2\pi$$

Therefore,

$$I = \pi$$

Preface

What is the value of this integral?

$$\int_0^1 x^2 \, \mathrm{d}x$$

Hopefully, you calculated the correct value of $\frac{1}{3}$. Now, what about this one?

$$\int_0^{\frac{\pi}{2}} \ln(\sin x) \, \mathrm{d}x$$

Well, that was quite a jump in difficulty. If your pencil is already out, trying out your every tool, then this book is for you! However, if you are perplexed as to why anyone *cares* about the evaluation of this integral, then this book is for you as well!

The complete solution to the integral above can be found in (2.2). However, the value has little importance. Rather, it is the *techniques* and *methods* that are employed that are worth attention. If problems like these give you a kick, then you are in for a good ride. If not, then by part 4 of this book, you will see the importance of the techniques used to answer such questions!

I have written this book with two types of readers in mind: 1) mathematical enthusiasts who love a challenging problem, and

2) physics, chemistry, and engineering majors. In this book, I aim to use the methods introduced in a standard two-semester calculus course to develop both problem solving skills in mathematics *and* the mathematical sciences.

The examples given often have very differing solutions, some of marvelous ingenuity and some that are rather standard. This is done on purpose, as it ultimately benefits readers to see multiple perspectives on similar problems. From u−substitutions to clever interchanges of integration and summation, various methods will be presented that can be used to solve the same problems. In order to make this book accessible to a larger base of students, contour integration is not included in the book.

It is worth noting that this is not an elementary calculus book, although the first chapters are there as a refresher for those who need it. A key difference between this book and other mathematics books that attempt to address a similar topic is that it is written *with the reader in mind*. Rarely would you need to struggle through proving a non-trivial statement that was previously declared as "trivial and left to the reader." Moreover, instead of the normal theorem-focused advanced mathematics book, I aim to minimize the number of techniques and methods and instead focus on *examples*.

In writing this book, I intended to make it as self-contained as possible. The various identities and theorems used in this book are often proved in the book, and the scientific concepts behind each application are explained and elaborated upon. Moreover, for theorems that need an extensive mathematics background, I aimed to simplify and translate their statements into the book's area of concern as best as possible without losing key details.

Even though this book is heavily mathematical, almost 100 pages are dedicated to applications of the techniques explored in the book. The applications are broad and include various topics of concern in the sciences and engineering. Many of the problems in this book, particularly those in chapter 8, are included simply because they are elegant results and develop the problem-solving skills of the reader. However, a few of the integrals and series, and certainly all of the methods employed, have wide applications in many science and engineering fields.

In the first two chapters, I introduce Mathematica as a way to numerically or symbolically verify the correctness of a result. This is done to familiarize the reader with Mathematica syntax so they are able to employ it on their own in later chapters.

In addition to almost one hundred fully worked out examples, there are exercise problems for the reader at the end of each chapter. Generally, these problems get progressively harder by number. However, all these problems involve the same techniques used in the chapter they are included in! A few challenge problems are scattered throughout the book to engage the more experienced reader.

The answers to these exercise problems are all included at the end of the book. There is a high likelihood that I will compile a list of solutions to these problems in a solutions manual to be published a few months after the book's launch.

Enjoy!

Hamza Alsamraee,
Centreville High School

A Note on the Originality of the Results

I have tried to cite results attributed to well-known mathematicians to the best of his ability. However, it is virtually impossible to check the originality of all the results in this book. Many are classic results, and a few are more unusual. To the best of my knowledge, many of the integrals and series in this book will be exposed to the literature for the first time. I do not claim originality, but I do claim authenticity.

Part I

Introductory Chapters

Chapter 1

Differential Calculus

This chapter will serve as a review of differential calculus, which will be used throughout the book, especially in chapter 3. In this chapter, we will also delve into elementary as well as advanced limits, giving a glimpse into the later chapters of the book. We will begin with the epsilon-delta definition of the limit and transition into the evaluation of limits. After all, what better way is there to start a calculus book other than to define the limit?

1.1 The Limit

We shall define the limit as the following (epsilon-delta definition):

> **Definition**
>
> Let $f(x)$ be a function defined on the interval (a,b), except possibly at x_0, where x_0 is in the interval i.e. $a < x_0 < b$ then the limit is $\lim\limits_{x \to x_0} f(x) = L$ if for every number $\varepsilon > 0$ there exists a $\delta > 0$ such that
>
> $$|f(x) - L| < \varepsilon \text{ whenever } 0 < |x - x_0| < \delta \quad (1.1)$$
>
> For all x.

This is a formalization of the limit which turns our rather informal notion of the limit to a rigorous one. Instead of using broad terms such as $f(x)$ gets "close" to L as x gets "close" to x_0, this definition allows us to rigorously discuss limits.

The theorem originated from the French mathematician and physicist Augustin-Louis Cauchy and was modernized by the German mathematician Karl Weierstrass. This marked an interesting time in the history of mathematics, representing its move towards rigor.

1.1. THE LIMIT

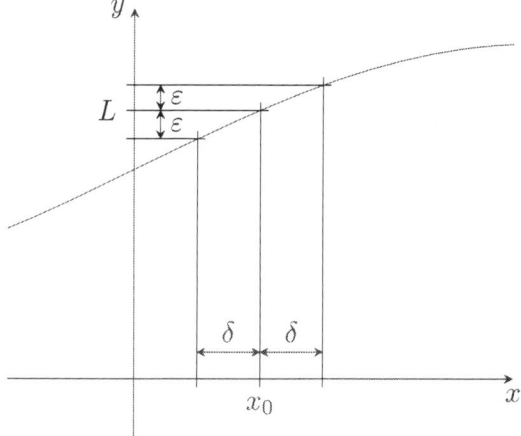

Figure 1.1: A visualization of the epsilon-delta definition of the limit

> "...Since Newton the limit had been thought of as a bound which could be approached closer and closer, though not surpassed. By 1800, with the work of L'Huilier and Lacroix on alternating series, the restriction that the limit be one-sided had been abandoned. Cauchy systematically translated this refined limit-concept into the algebra of inequalities, and used it in proofs once it had been so translated; thus he gave reality to the oft-repeated eighteenth-century statement that the calculus could be based on limits." - American mathematician Judith Grabiner[a]
>
> [a]Grabiner, Judith V. (March 1983), "Who Gave You the Epsilon? Cauchy and the Origins of Rigorous Calculus", The American Mathematical Monthly, 90 (3): 185–194, doi:10.2307/2975545, JSTOR 2975545

This formulation will not be used extensively in this book, but is nonetheless widely used in analysis. Specifically, it is employed heavily in proving the continuity of a function.

> **Definition**
>
> A function f is said to be continuous at x_0 if it is both defined at x_0 and its value at x_0 equals the limit of $f(x)$ as x approaches x_0, i.e.
>
> $$\lim_{x \to x_0} f(x) = f(x_0)$$

Consequently, $f(x)$ is said to be continuous on some interval (a, b) if it is continuous for every x_0 belonging to that interval.

Let us begin with an easy first example!

Example 1: Prove that $\lim_{x \to 0} x^2 = 0$ using the epsilon-delta definition of a limit.

Solution

In this case both L and x_0 are zero. We start by letting $\varepsilon > 0$. According to the (ε, δ) definition of the limit, if $\lim_{x \to 0} x^2 = 0$ we will need to find some other number $\delta > 0$ such that

$$|x^2 - 0| < \varepsilon \text{ whenever } 0 < |x - 0| < \delta$$

Which gives us

$$x^2 < \varepsilon \text{ whenever } 0 < |x| < \delta$$

Starting with the left inequality and taking the square root of both sides we get

$$|x| < \sqrt{\varepsilon}$$

This looks very similar to the right inequality, which drives us to set $\sqrt{\varepsilon} = \delta$. We now need to prove that our choice satisfies

$$|x^2| < \varepsilon \text{ whenever } 0 < |x| < \sqrt{\varepsilon}$$

1.1. THE LIMIT

Starting with the right inequality with the assumption that $0 < |x| < \sqrt{\varepsilon}$

$$\left|x^2\right| = |x|^2 < \left(\sqrt{\varepsilon}\right)^2 = \varepsilon$$

Which gives us

$$\left|x^2\right| < \varepsilon$$

This is exactly what we needed to show! In conclusion, we have shown that for any $\varepsilon > 0$ we can find a $\delta > 0$ such that

$$\left|x^2 - 0\right| < \varepsilon \text{ whenever } 0 < |x - 0| < \delta$$

Which proves our original supposition that

$$\lim_{x \to 0} x^2 = 0$$

Example 2: Prove that $\lim_{x \to 0} \dfrac{\sin(x)}{x} = 1$ using the epsilon-delta definition of a limit.

Solution

To start off, we will make use of the unit circle to derive some identities. Consider figure 1.2.

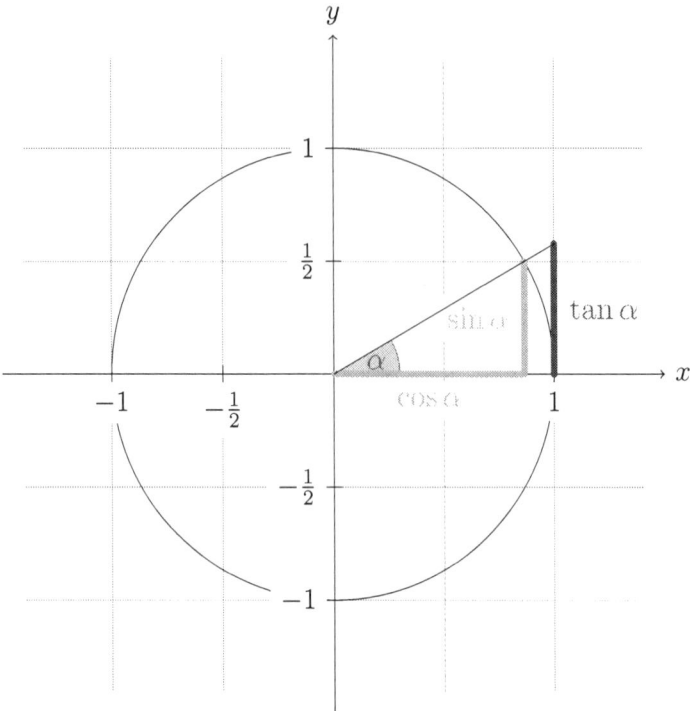

Figure 1.2: The unit circle. Figure generated using TikZ software

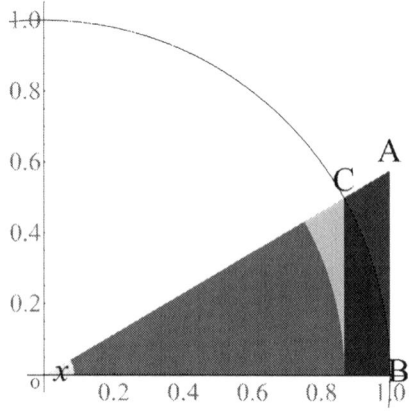

Figure 1.3: Points B, C lie on the unit circle

1.1. THE LIMIT

Now, consider figure 1.3. Using the basics of the unit circle, we can say that point C has a y-coordinate equal to $\sin x$ and point A has a y-coordinate equal to $\tan x$. Now, consider triangle $\triangle OCB$ and triangle $\triangle OAB$ where O is the origin. $\triangle OCB$ has area is $\frac{1}{2}\sin x$ while $\triangle OAB$ has area $\frac{1}{2}\tan x$.

Moreover, the area of the sector formed by x is $\frac{x}{2}$ (x is measured in radians). We now have the inequality

$$\frac{1}{2}\sin x \leq \frac{x}{2} \leq \frac{1}{2}\tan x \tag{1.2}$$

Dividing (1.2) by $\frac{1}{2}\sin x$,

$$1 \leq \frac{x}{\sin x} \leq \frac{1}{\cos x}$$

Taking the reciprocal,

$$\cos x \leq \frac{\sin x}{x} \leq 1$$

Since $\frac{\sin x}{x}$ and $\cos x$ are even functions, this inequality holds for any non-zero x on the interval $\left(-\frac{\pi}{2}, \frac{\pi}{2}\right)$. This leads us to

$$\left|\frac{\sin(x)}{x} - 1\right| < 1 - \cos(x) \tag{1.3}$$

We can then use the trigonometric identity

$$\cos(2x) = 1 - 2\sin^2 x$$

To obtain that $1 - \cos(x) = 2\sin^2\left(\frac{x}{2}\right) < \frac{x^2}{2}$ where the inequality holds because of (1.2). (1.3) is then transformed into

$$\left|\frac{\sin(x)}{x} - 1\right| < \frac{x^2}{2} \tag{1.4}$$

We proceed to let $\delta = \sqrt{\varepsilon}$ which gives us

$$|x - 0| < \delta = \sqrt{\varepsilon}$$

Plugging back into (1.4),

$$\left|\frac{\sin(x)}{x} - 1\right| < \frac{x^2}{2} < \frac{\sqrt{\varepsilon^2}}{2} < \varepsilon$$

Which then proves the limit by showing that for any $\varepsilon > 0$ we can find a $\delta > 0$ such that

$$\left|\frac{\sin(x)}{x} - 1\right| < \varepsilon \text{ whenever } 0 < |x - 0| < \delta$$

1.1.1 L'Hopital's Rule

> **Theorem**
>
> L'Hopital's rule is one of the most widely known limit properties. It states that if two functions f and g are differentiable on an open interval I, with an exception made for a, and the limit of their quotient takes an indeterminate form such as $\frac{0}{0}$ or $\frac{\infty}{\infty}$, the limit of their quotient can be expressed as
>
> $$\lim_{x \to a} \frac{f(x)}{g(x)} = \lim_{x \to a} \frac{f'(x)}{g'(x)}$$
>
> Given that $g'(x) \neq 0$ and the RHS exists.

The rule is named after the 17th-century French mathematician Guillaume de L'Hopital. Even though the rule is often attributed to L'Hopital, the theorem was first introduced to him in 1694 by the Swiss mathematician Johann Bernoulli. This rule simplifies many limits and sometimes is needed more than once. Again, we can start with an easy problem to get the gist of how to use L'Hopital's rule.

1.1. THE LIMIT

Example 3: Find the value of $\lim\limits_{x \to \infty} xe^{-x}$

Solution

We will first need to establish a non-constant denominator. This is because if our denominator is constant, its derivative will be 0. There are two simple choices

$$\frac{e^{-x}}{\frac{1}{x}}$$

Or

$$\frac{x}{e^x}$$

We can see that the latter choice is better as the first will produce an even more complicated result when L'Hopital's rule is applied. Remember, when one key does not work, one must proceed to try another one! As the problems in this book get harder, this idea becomes vital. Going back to our problem,

$$L = \lim_{x \to \infty} xe^{-x} = \lim_{x \to \infty} \frac{x}{e^x}$$

After applying L'Hopital's rule we obtain:

$$L = \lim_{x \to \infty} \frac{1}{e^x}$$

Now it is easy to see that

$$L = \lim_{x \to \infty} xe^{-x} = 0$$

To numerically "confirm" our result, we can use Wolfram Desktop (Or Mathematica):

In[1]:= `Limit[x/E^x, x -> Infinity]`

Out[1]= 0

Example 4: Find $\lim\limits_{n\to\infty} \left[\dfrac{\left(1+\frac{1}{n}\right)^n}{e}\right]^n$

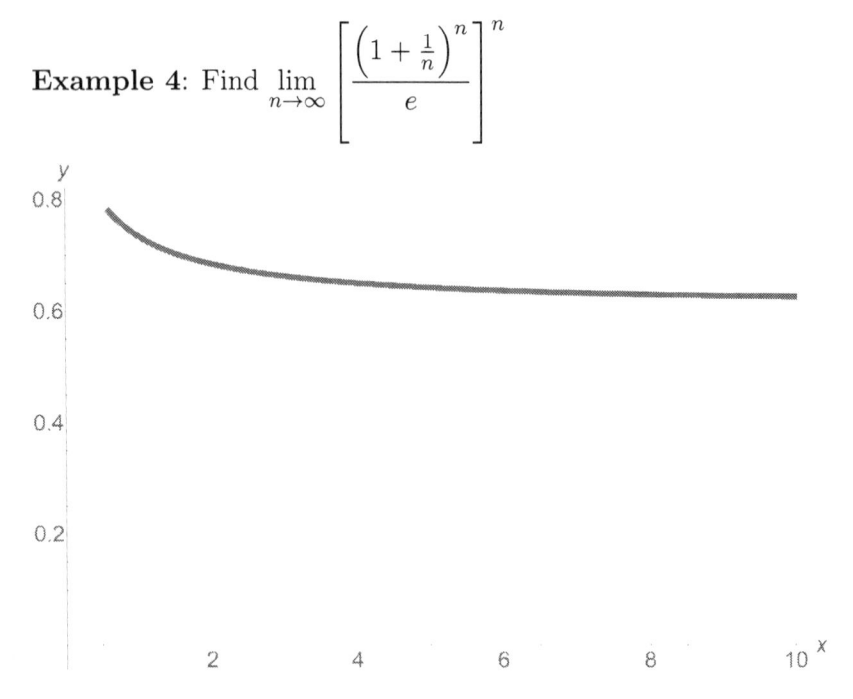

Figure 1.4: The graph shows limiting behavior approaching a value near .6

Solution
Define
$$L = \lim_{n\to\infty} \left[\dfrac{\left(1+\frac{1}{n}\right)^n}{e}\right]^n$$

We then have

$$\ln L = \lim_{n\to\infty} n\ln\left(\dfrac{\left(1+\frac{1}{n}\right)^n}{e}\right) = \lim_{n\to\infty}\left[n^2\ln\left(1+\dfrac{1}{n}\right) - n\right]$$

We will proceed to transform the limit into one we can apply L'Hopital's rule on, i.e. a fraction with a non-constant denomi-

1.1. THE LIMIT

nator:
$$\ln L = \lim_{n \to \infty} \frac{n \ln \left(1 + \frac{1}{n}\right) - 1}{\frac{1}{n}}$$

After applying L'Hopital's we obtain:

$$\ln L = \lim_{n \to \infty} \frac{\ln\left(1+\frac{1}{n}\right) - \frac{n \cdot \frac{1}{n^2}}{1+\frac{1}{n}}}{-\frac{1}{n^2}} = \lim_{n \to \infty} \frac{\ln\left(1+\frac{1}{n}\right) - \frac{1}{n+1}}{-\frac{1}{n^2}}$$

Applying L'Hopital's rule again and simplifying we obtain:

$$\ln L = \lim_{n \to \infty} \frac{-\frac{1}{n(n+1)^2}}{\frac{2}{n^3}} = -\lim_{n \to \infty} \frac{n^3}{2n(n+1)^2}$$

We can already see that expanding the denominator would result in a first term of $2n^3$. Thus,

$$\ln L = -\frac{1}{2}$$

By using the ratio of the coefficients.

$$\implies L = \frac{1}{\sqrt{e}}$$

Mathematica gives:

In[2]:= `Limit[((1 + 1/x)^x/E)^x, x -> Infinity]`

Out[2]= `1/Sqrt[E]`

1.1.2 More Advanced Limits

Example 5: Define the Riemann zeta function as $\zeta(s) = \sum_{n=1}^{\infty} \frac{1}{n^s} = \frac{1}{1^s} + \frac{1}{2^s} + \frac{1}{3^s} + \frac{1}{4^s} \ldots$ Find $\lim_{s \to \infty} (\zeta(s) - 1)^{\frac{1}{s}}$

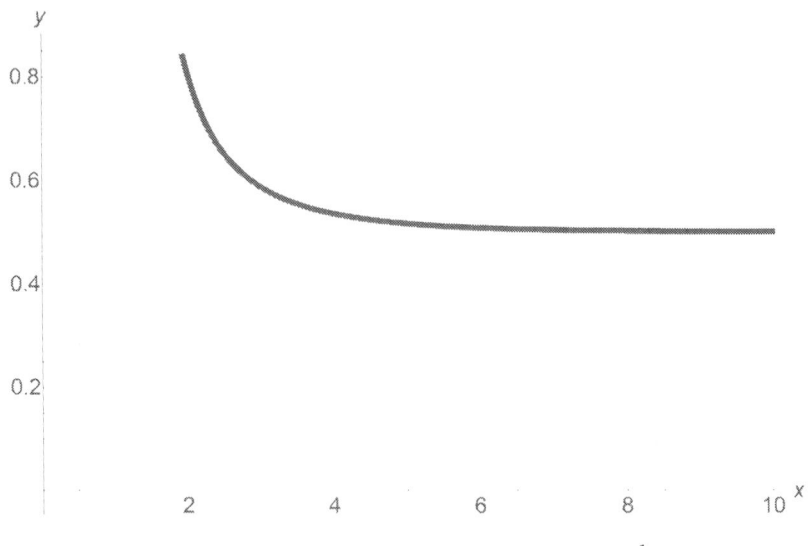

Figure 1.5: A graph of $y = \left(\zeta(x) - 1\right)^{\frac{1}{x}}$

Solution

We know that the first term of the zeta function will always be

$$\frac{1}{1^s} = 1$$

Therefore,

$$L = \lim_{s \to \infty} \left(\zeta(s) - 1\right)^{\frac{1}{s}}$$

$$= \lim_{s \to \infty} \left(1 + \frac{1}{2^s} + \frac{1}{3^s} + \frac{1}{4^s} \cdots - 1\right)^{\frac{1}{s}}$$

$$= \lim_{s \to \infty} \left(\frac{1}{2^s} + \frac{1}{3^s} + \frac{1}{4^s} \cdots\right)^{\frac{1}{s}}$$

We proceed to factor out $\dfrac{1}{2^s}$

$$L = \lim_{s \to \infty} \frac{1}{2} \left(1 + \left(\frac{2}{3}\right)^s + \left(\frac{2}{4}\right)^s \cdots\right)^{\frac{1}{s}}$$

1.1. THE LIMIT

As $s \to \infty$, all terms inside the parentheses with s as an exponent will approach zero. Therefore,

$$L = \lim_{s \to \infty} \frac{1}{2} \cdot 1^{\frac{1}{s}}$$

Now it is easy to see that

$$\lim_{s \to \infty} \left(\zeta(s) - 1\right)^{\frac{1}{s}} = \frac{1}{2}$$

Testing our result with Mathematica,

In[3]:=   (Zeta[x]-1)^(1/x)

Out[3]= $\dfrac{1}{2}$

Example 6: Define the double factorial, which is usually denoted !!, as

$$n!! = n \cdot (n-2) \cdot (n-4) \cdots$$

For even numbers, the last number to be multiplied by is 2, while for odd numbers it is 1. For example,

$$6!! = 6 \cdot 4 \cdot 2 = 48$$

$$5!! = 5 \cdot 3 \cdot 1 = 15$$

Find $\displaystyle\lim_{n \to \infty} \frac{(2n-1)!! \sqrt{n}}{(2n)!!}$ for $n \in \mathbb{N}$

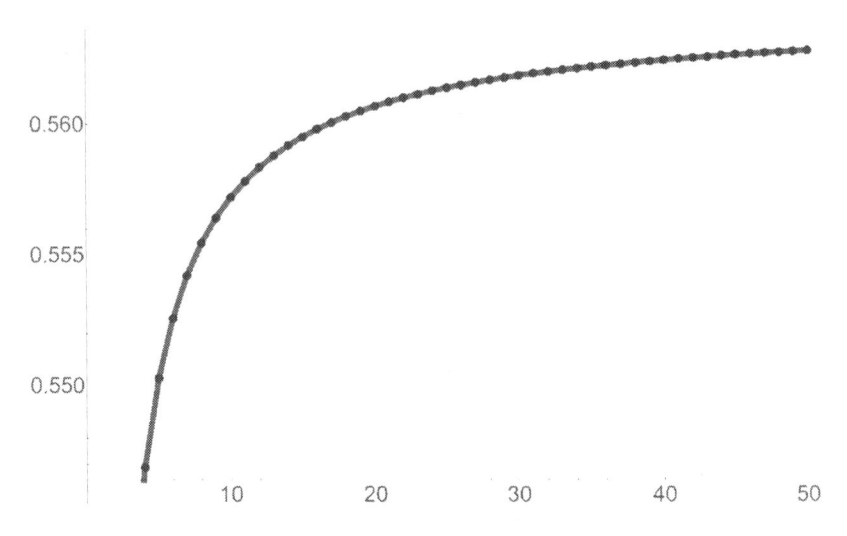

Figure 1.6: Using Mathematica's ListLinePlot function, we can get a graph of the sequence $a_n = \frac{(2n-1)!!\sqrt{n}}{(2n)!!}$

Solution
A double factorial, $n!!$, can be expressed using normal factorials. The expressions for our case are as follows

$$(2n)!! = 2^n \cdot n!$$

$$(2n-1)!! = \frac{(2n)!}{2^n \cdot n!}$$

By substituting the above expressions into our limit we obtain:

$$L = \lim_{n \to \infty} \frac{(2n)!\sqrt{n}}{(n)!^2 \cdot 4^n}$$

Now, we will introduce Stirling's approximation.

1.1. THE LIMIT

> **Theorem**
>
> Stirling's approximation, or Stirling's formula, is one of the most common approximations for factorials. The general formula is
>
> $$\ln k! = k \ln k - k + O(\ln k) \qquad (1.5)$$
>
> Where Big O notation indicates that the LHS (Left-hand side) describes the RHS's (Right-hand side) limiting behavior, i.e. as $k \to \infty$. A more precise approximation, which is derived from (1.5), is
>
> $$n! \sim \sqrt{2\pi n} \left(\frac{n}{e}\right)^n \qquad (1.6)$$

Using (1.6) we have:

$$L = \lim_{n \to \infty} \frac{\sqrt{4\pi n}(2n)^{2n} \cdot e^{2n}}{e^{2n} \cdot (2\pi n) \cdot n^{2n} \cdot 2^{2n}} \sqrt{n}$$

After simplifying, we are only left with

$$L = \lim_{n \to \infty} \frac{\sqrt{n \cdot 4\pi n}}{2\pi n}$$

$$= \lim_{n \to \infty} \frac{2n\sqrt{\pi}}{2\pi n}$$

$$\implies L = \frac{1}{\sqrt{\pi}} \approx .56419$$

We can test our result with Mathematica. Unfortunately, Mathematica is unable to predict a value due to the computational heaviness of the factorial. However, evaluating our sequence for $n = 1,000,000$ gives us $\approx .56419$, so we can be confident in our result.

Example 7: Define the nth Fibonacci number as $F_n = F_{n-1} + F_{n-2}$ where $F_1 = F_2 = 1$. Find $\lim\limits_{n \to \infty} \dfrac{F_{n+1}}{F_n}$.

Solution
We start by noting that through the definition of the Fibonacci sequence we have

$$F_{n+1} = F_n + F_{n-1} \implies \frac{F_{n+1}}{F_n} = 1 + \frac{F_{n-1}}{F_n}$$

Define
$$a_n = \frac{F_{n+1}}{F_n}$$

Then,

$$a_n = 1 + \frac{1}{a_{n-1}}$$

Taking the limit of both sides,

$$\lim_{n \to \infty} a_n = \lim_{n \to \infty} \left(1 + \frac{1}{a_{n-1}}\right)$$

Since both the limits of the RHS and LHS exist we can let $\lim_{n \to \infty} a_n = x$ to get

$$x = 1 + \frac{1}{x}$$
$$x^2 - x - 1 = 0$$

Using the quadratic formula and eliminating the negative solution we obtain that

$$\lim_{n \to \infty} \frac{F_{n+1}}{F_n} = \frac{1 + \sqrt{5}}{2} = \phi$$

Where ϕ denotes the famous golden ratio. To check our result, we can use Mathematica's built in Fibonacci sequence function:

1.2. THE DERIVATIVE

In[4]:= `Limit[Fibonacci[n+1]/Fibonacci[n],n->Infinity]`

Out[4]= $\frac{1}{2} (1+\sqrt{5})$

1.2 The Derivative

> **Definition**
>
> The derivative of a function $f(x)$, denoted by $f'(x)$, gives us the instantaneous rate of change at a point $(x, f(x))$ by using the concept of a limit. This is done by measuring the "slope" over an infinitesimal interval, i.e. $[x, x+h]$.
>
> $$f'(x) = \lim_{h \to 0} \frac{f(x+h) - f(x)}{h}$$

Example 8: What is the derivative of the function $f(x) = x^n$, where n is a constant?

Solution

Many have already memorized the power rule, but let us prove it here. Using the definition of a derivative,

$$\frac{d}{dx} f(x) = \lim_{h \to 0} \frac{(x+h)^n - x^n}{h}$$

We now recall the binomial theorem which states that

$$(x+a)^n = \sum_{k=0}^{n} \binom{n}{k} x^{n-k} a^k$$

Therefore,

$$\frac{\mathrm{d}}{\mathrm{d}x} f(x) = \lim_{h \to 0} \frac{\sum_{k=0}^{n} \binom{n}{k} x^{n-k} h^k - x^n}{h}$$

By expanding this expression, we can see that the first term, x^n, cancels out and we are left with

$$\frac{\mathrm{d}}{\mathrm{d}x} f(x) = \lim_{h \to 0} \frac{nx^{n-1}h + \frac{n(n-1)}{2!} x^{n-2} h^2 + \dots}{h}$$

Which simplifies to

$$\frac{\mathrm{d}}{\mathrm{d}x} f(x) = \lim_{h \to 0} nx^{n-1} + \frac{n(n-1)}{2!} x^{n-2} h^1 + \dots$$

Notice that the degree of h increases. As $h \to 0$, all terms but the first approach 0. Thus,

$$\frac{\mathrm{d}}{\mathrm{d}x} f(x) = nx^{n-1}$$

1.2.1 Product Rule

> **Theorem**
>
> For two differentiable functions, the derivative of their product can be stated as follows
>
> $$\frac{\mathrm{d}}{\mathrm{d}x} uv = u\frac{\mathrm{d}}{\mathrm{d}x} v + v\frac{\mathrm{d}}{\mathrm{d}x} u$$

1.2. THE DERIVATIVE

Example 9: Find $\lim\limits_{x \to 0} \left(\dfrac{x}{\sin^3 x} - \dfrac{1}{x^2} \right)$.

Solution

The easiest way to go about this problem is using Maclaurin series, however we will save that for later chapters. We will attempt a more basic approach by factoring the above expression.

$$L = \lim_{x \to 0} \left(\frac{x}{\sin^3 x} - \frac{1}{x^2} \right)$$

$$= \lim_{x \to 0} \left(\frac{x^3 - \sin^3 x}{x^2 \sin^3 x} \right)$$

Now we will attempt to convert this into a product of multiple limits. It is easy to see that $x^3 - \sin^3 x = (x - \sin x)(x^2 + x \sin x + \sin^2 x)$. Hence,

$$L = \lim_{x \to 0} \left(\frac{x - \sin x}{\sin^3 x} \right) \cdot \lim_{x \to 0} \left(\frac{x^2 + x \sin x + \sin^2 x}{x^2} \right)$$

Both limits are $\frac{0}{0}$ cases so we will proceed to apply L'Hopital's rule.

$$L = \lim_{x \to 0} \left(\frac{1 - \cos x}{3 \sin^2 x \cos x} \right) \cdot \lim_{x \to 0} \left(\frac{2x + \sin x + x \cos x + 2 \sin x \cos x}{2x} \right)$$

Applying the rule again to the first limit,

$$L = \lim_{x \to 0} \left(\frac{\sin x}{6 \sin x \cos^2 x - 3 \sin^3 x} \right)$$
$$\cdot \lim_{x \to 0} \left(\frac{2x + \sin x + x \cos x + 2 \sin x \cos x}{2x} \right)$$

One last round to both limits!

$$L = \lim_{x \to 0} \left(\frac{\cos x}{6\cos^3 x - 21\sin^2 x \cos x} \right)$$
$$\cdot \lim_{x \to 0} \left(\frac{2 + 2\cos x - x\sin x + 2(\cos^2 x - \sin^2 x)}{2} \right)$$

Now the limit is simplified to an easy plug-in. Our limit is therefore:

$$L = \frac{1}{6} \cdot 3 = \frac{1}{2}$$

Evaluating our limit using Mathematica,

In[5]:= `Limit[x/Sin[x]^3 - 1/x^2, x -> 0]`

Out[5]= $\frac{1}{2}$

1.2.2 Quotient Rule

> **Theorem**
>
> Let $f(x) = \frac{g(x)}{h(x)}$ where both h and g are differentiable and $h(x) \neq 0$. The derivative of $f(x)$ is then
>
> $$f'(x) = \frac{g'(x)h(x) - h'(x)g(x)}{[h(x)]^2}$$

This rule can be found from the product rule by setting $u = g(x)$ and $v = \frac{1}{h(x)}$. However, it is helpful to know on its own.

Example 10: Find $\frac{d}{dx} \tan x$

1.2. THE DERIVATIVE

Solution

We know that
$$\tan x = \frac{\sin x}{\cos x}$$

Applying the quotient rule,

$$(\tan x)' = \frac{(\sin x)' \cos x - (\cos x)' \sin x}{\cos^2 x}$$

$$= \frac{\cos^2 x + \sin^2 x}{\cos^2 x}$$

$$= \frac{1}{\cos^2 x}$$

We then have by the definition of $\sec x$,

$$\frac{d}{dx} \tan x = \sec^2 x$$

1.2.3 Chain Rule

Theorem

If $g(x)$ is differentiable at $x = a$ and $f(x)$ is differentiable at $x = g(a)$, then the derivative of $f(g(x))$ at $x = a$ is:

$$\frac{d}{dx} f(g(x)) \bigg|_{x=a} = f'(g(a)) \cdot g'(a)$$

After a basic review, we now can start delving into more advanced derivatives. Let us get started with an interesting derivative!

Example 11: Find the derivative of $f(x) = x^{x^x}$ at $x = 1$.

Solution

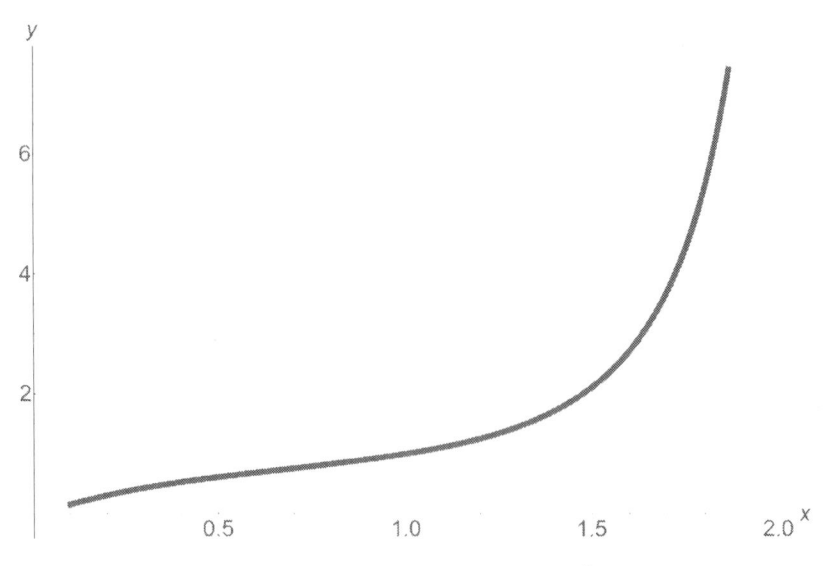

Figure 1.7: Graph of $y = x^{x^x}$

We will approach this problem using *logarithmic differentiation*. We first set $h(x) = x^x$ and then take the natural logarithm of both sides,
$$\ln h(x) = x \ln x$$
Differentiating both sides using the chain rule,
$$\frac{h'(x)}{h(x)} = \ln x + 1 \to h'(x) = x^x (\ln x + 1)$$
We can now return to our original function and take the natural logarithm of both sides to get:
$$\ln f(x) = x^x \ln x$$
Differentiating both sides,
$$\frac{f'(x)}{f(x)} = \frac{\mathrm{d}}{\mathrm{d}x}(x^x) \ln x + x^{x-1}$$
$$= x^x \left(\frac{1}{x} + \ln^2 x + \ln x \right)$$

1.2. THE DERIVATIVE

Plugging in $x = 1$,

$$\frac{f'(1)}{1} = 1 \to f'(1) = 1$$

Checking with Mathematica,

```
In[6]:= f[x_]:=x^x^x
        f'[1]

Out[6]=  1
```

Example 12: Evaluate $\dfrac{d^\pi}{dx^\pi} x^\pi$ at $x = \pi$

Solution

You might be wondering what a πth derivative even is, or if it is even valid mathematically! This notion of generalizing the derivative operator is the basis of an entire branch of mathematics, **fractional calculus**. The first appearance of a fractional derivative is in a letter written to l'Hôpital by the German polymath Gottfried Wilhelm Leibniz in 1695[1]. The theory and foundations of the subject were introduced by the likes of the French mathematician Joseph Liouville in the 19th century. In the late 19th century, the electrical engineer Oliver Heaviside introduced their practical use in electrical transmission line analysis[2].

To solve this problem, let's look at the function $f(x) = x^a$. By the power rule, we can deduce that the first derivative is

$$f'(x) = ax^{a-1}$$

---

[1] Katugampola, Udita N. (15 October 2014). *A New Approach To Generalized Fractional Derivatives*. Bulletin of Mathematical Analysis and Applications. 6 (4): 1–15. arXiv:1106.0965. Bibcode:2011arXiv1106.0965K.

[2] Bertram Ross (1977). *The development of fractional calculus 1695-1900*. Historia Mathematica. 4: 75–89. doi:10.1016/0315-0860(77)90039-8

CHAPTER 1. DIFFERENTIAL CALCULUS

Applying differentiation repeatedly gives us a pattern, namely that the nth derivative is

$$\frac{\mathrm{d}^n}{\mathrm{d}x^n} x^a = \frac{a!}{(a-n)!} x^{a-n}$$

We can extend this definition using the Gamma function.

> **Definition**
>
> The Gamma function, usually denoted by $\Gamma(\cdot)$, extends the domain of the factorial function into $\mathbb{C}$ with the exception of non-positive integers.
>
> $$\Gamma(n) = (n-1)!$$
>
> It is mainly defined by a convergent improper integral. However, the product definition due to Weirstrass is also used[a].
>
> $$\Gamma(z) = \int_0^\infty t^{z-1} e^{-t} \mathrm{d}t = \frac{e^{-\gamma z}}{z} \prod_{k=1}^\infty \left(1 + \frac{z}{k}\right)^{-1} e^{z/k} \quad (1.7)$$
>
> ---
> [a]Davis, P. J. (1959). *Leonhard Euler's Integral: A Historical Profile of the Gamma Function"*. American Mathematical Monthly. 66 (10): 849–869. doi:10.2307/2309786. JSTOR 2309786

1.2. THE DERIVATIVE

> **Definition**
>
> The Euler-Mascheroni constant, usually denoted as γ, is defined as the limiting difference between the harmonic sum and the natural logarithm:
>
> $$\gamma = \lim_{k \to \infty} \left(-\ln k + \sum_{n=1}^{k} \frac{1}{n} \right) \approx 0.577216 \qquad (1.8)$$
>
> Note: The partial harmonic sum $\sum_{n=1}^{k} \frac{1}{n}$ is sometimes also written as H_k (Harmonic numbers).

Since the argument of the Gamma function is shifted down by 1, we can rewrite the nth derivative as

$$\frac{d^n}{dx^n} x^a = \frac{\Gamma(a+1)}{\Gamma(a-n+1)} x^{a-n}$$

After plugging in $a = n = \pi$ we get:

$$\frac{d^\pi}{dx^\pi} x^\pi = \frac{\Gamma(\pi+1)}{\Gamma(1)} x^0 = \Gamma(\pi+1) \approx 7.188$$

Where we used Mathematica to obtain the numerical approximation. Unfortunately, Mathematica has no built in fractional derivative function to date.

Example 13: Let $f(x) = \dfrac{x}{x + \frac{x}{x+\frac{x}{x+..}}}$. Find $\lim_{x \to \infty} f'(x)$.

Solution
We will first find an expression for the derivative and then take the limit. Notice that

$$f(x) = y = \frac{x}{x + \frac{x}{x+\frac{x}{x+..}}} = \frac{x}{x+y}$$

Therefore,
$$xy + y^2 = x$$
$$y^2 + xy - x = 0$$

This is a quadratic equation in y. By the quadratic formula we obtain:
$$y = \frac{-x \pm \sqrt{x^2 + 4x}}{2}$$

Since we know that $y > 0$ for all $x > 0$ we will ignore the negative value. Hence,
$$y = \frac{-x + \sqrt{x^2 + 4x}}{2}$$

To get our desired derivative expression we will differentiate to get:
$$\frac{dy}{dx} = \frac{1}{2}\left(\frac{2x+4}{2\sqrt{x^2+4x}} - 1\right)$$
$$= \frac{1}{2}\left(\frac{x+2}{\sqrt{x^2+4x}} - 1\right)$$

Taking the limit we obtain:
$$\lim_{x \to \infty} f'(x) = \frac{1}{2} \lim_{x \to \infty} \left(\frac{x+2}{\sqrt{x^2+4x}} - 1\right)$$
$$= \frac{1}{2} \lim_{x \to \infty} \left(\frac{x+2}{\sqrt{x^2+4x}}\right) - \frac{1}{2}$$
$$= \frac{1}{2} \lim_{x \to \infty} \left(\frac{x+2}{\sqrt{(x+2)^2 - 4}}\right) - \frac{1}{2}$$

We can see that the denominator tends to $x+2$ as $x \to \infty$. Our limit is then simply
$$\lim_{x \to \infty} f'(x) = \frac{1}{2} - \frac{1}{2} = 0$$

1.2. THE DERIVATIVE

Example 14: Find $\frac{d^2x}{dy^2}$ in terms of $\frac{d^2y}{dx^2}$ and $\frac{dy}{dx}$.

Solution
Define
$$g(x) = f^{-1}(x)$$
With $g(x) = y$. Then by the basic properties of inverse functions we have
$$f(g(x)) = x$$
Differentiating both sides with respect to x and using the chain rule,
$$f'(g(x)) \cdot g'(x) = 1$$
$$f'(g(x)) = \frac{1}{g'(x)}$$
Differentiating once again we get:
$$f''(g(x)) \cdot g'(x) = -\frac{g''(x)}{(g'(x))^2}$$
$$\therefore f''(g(x)) = -\frac{g''(x)}{(g'(x))^3}$$
Expressing this in different notation,
$$\frac{d^2x}{dy^2} = -\frac{\frac{d^2y}{dx^2}}{\left(\frac{dy}{dx}\right)^3}$$

Example 15: Let $a > 1$ be a constant. What is the minimum value of a where the curves $f(x) = a^x$ and $g(x) = \log_a x$ are tangent?

Solution

We know that f and g are inverses, which means they are reflections of each other on the line $y = x$. Therefore, if they were to be tangent their derivatives have to be both 1. We then have a system of equations

$$a^{x_1} \ln a = \frac{1}{x_1 \ln a} = 1 \qquad (1)$$

$$a^{x_1} = \log_a x_1 \qquad (2)$$

Where (x_1, x_1) is the point of intersection. (1) and (2) are true since $f' = g' = 1$ at (x_1, x_1) and $f = g$ at (x_1, x_1), respectively. Substituting (2) into (1) we obtain:

$$\log_a x_1 \ln a = 1$$

Using base conversion,
$$\ln x_1 = 1$$
$$x_1 = e$$

Thus,
$$\frac{1}{e \ln a} = 1$$
$$e \ln a = 1$$
$$a = e^{\frac{1}{e}}$$

1.3 Exercise Problems

1) Prove the product rule using the limit definition of the derivative.

1.3. EXERCISE PROBLEMS

2) Prove the chain rule using the limit definition of the derivative.

3) Prove the quotient rule using logarithmic differentiation (See Example 11).

4) Find $\lim_{n \to \infty} n^{1/n}$

5) Evaluate $\lim_{x \to 0^+} x^x$

6) Find $\lim_{x \to 0^+} x^{x^x} - x^x$

7) Evaluate $\lim_{x \to 0} \dfrac{x^2 \sin\left(\frac{1}{x}\right) + x}{\sqrt[x]{1+x} - e}$

8) Let $f'(x) = f^{-1}(x)$. If $f(2019) = 2019$, what is $f''(2019)$?

9) Evaluate $\lim_{k \to \infty} \dfrac{\ln\left(\frac{k^k}{k!}\right)}{k}$

10) Evaluate $\lim_{n \to \infty} \left(\dfrac{\binom{4n}{3n}}{\binom{2n}{n}} \right)^{1/n}$ (Bonus question: Generalize it!)

11) Find $\lim_{x \to \infty} \dfrac{x^2}{2x+1} \sin\left(\dfrac{\pi}{x}\right)$ (Hint: See example 2 and use the squeeze theorem)

> **Theorem**
>
> Let (a, b) be an interval such that $x_0 \in (a, b)$. Let g, f, and h be functions defined on (a, b), except possibly at x_0. Suppose that for every $x \in (a, b)$ not equal to x_0, we have:
>
> $$g(x) \leq f(x) \leq h(x)$$
>
> Also, suppose that
>
> $$\lim_{x \to x_0} g(x) = \lim_{x \to x_0} h(x) = L$$
>
> Then
>
> $$\lim_{x \to x_0} f(x) = L$$

Chapter 2

Basic Integration

In this chapter, we will explore and review basic integration techniques such as u-substitution, integration by parts, partial fraction decomposition, etc. We will try to apply these methods creatively to a collection of elementary yet unique problems. With that in mind, let's get started!

$$\int$$

Figure 2.1: The notation for the indefinite integral, $\int$, was introduced by Leibniz in 1675. He adapted it from an archaic form of the letter s, standing for summa (Latin for "sum" or "total").

2.1 Riemann Integral

> **Definition**
>
> For a function $f(x)$ that is continuous on the interval $[a, b]$, we can divide the interval into n subintervals, each with length Δx, and from each interval choose a point x_k. The definite integral is then
>
> $$\int_a^b f(x)\ \mathrm{d}x = \lim_{n \to \infty} \sum_{k=1}^{n} f(x_k)\Delta x$$

This definition, provided by the German mathematician Bernhard Riemann, expresses area as a combination of infinitely many vertically-oriented rectangles, a technique known as Riemann sums. As the rectangles get thinner ($\Delta x \to 0$), the total

2.1. RIEMANN INTEGRAL

area of the rectangles approaches the value of the integral.

Perhaps the most advantageous aspect of the Riemann definition is its easy visualization! Consider the graph of $y = x^2$ in figure 2.2. We will begin by estimating the integral, or area un-

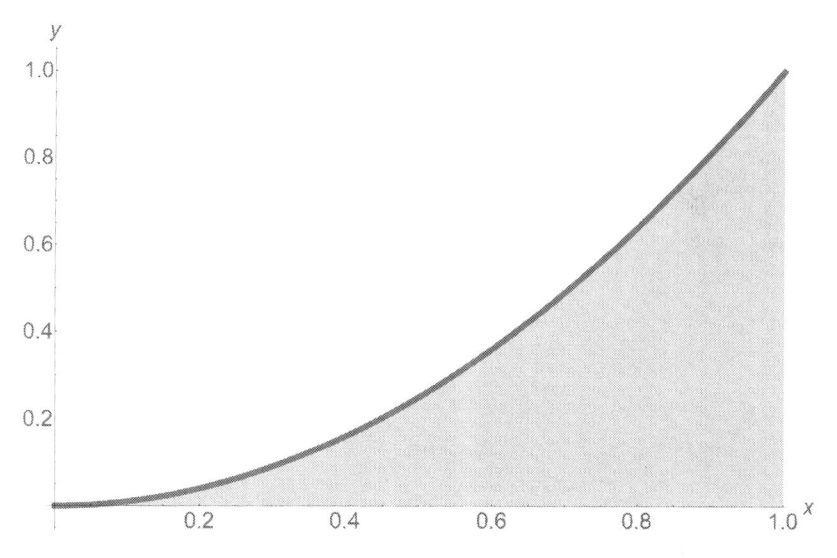

Figure 2.2: Graph of $y = x^2$ on $x \in [0, 1]$

der the curve. Since it is a nonstandard shape, we will approximate it using standard shapes, namely rectangles. We will start our approximation with two *subintervals*, or two rectangles.

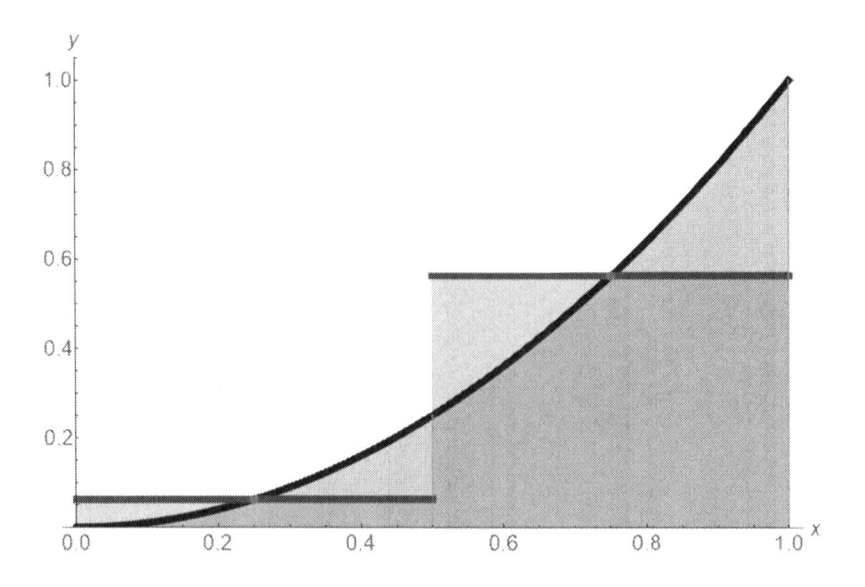

Figure 2.3: Graph of the midpoint Riemann sums for $y = x^2$ on $[0, 1]$ with only two subintervals

Note that this is a *midpoint* Riemann sum. Our thought experiment here will work for any Riemann sum, since as the rectangles get thinner and thinner, the difference between $f(x_k)$ of different Riemann sums will approach 0.

Now, back to our thought experiment. We can see that there are areas that our rectangles do not cover and some areas that our rectangles add. To minimize these errors, we have to add more rectangles! Let us now increase the number of subintervals from 2 to 10.

As one can see in figure 2.4, we now have a much better estimate of integral, or the area under the curve. As we let the number of subintervals approach ∞ (and consequently their width approach 0), this approximation will become an exact value for area!

2.2. LEBESGUE INTEGRAL

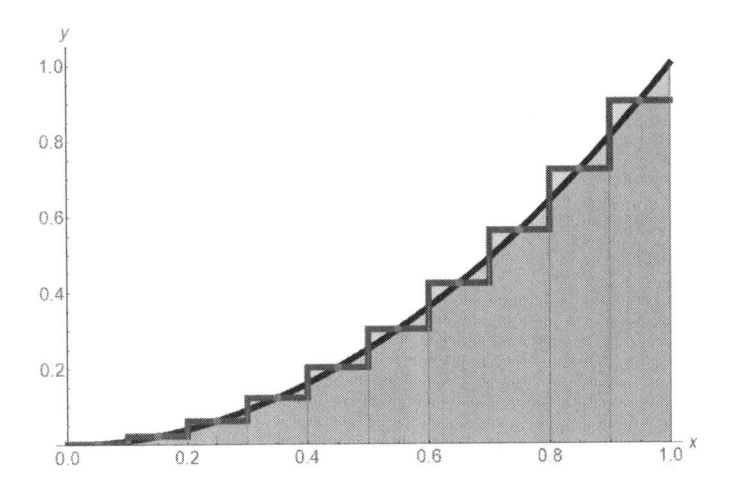

Figure 2.4: Graph of the midpoint Riemann sums for $y = x^2$ on $[0, 1]$ with ten subintervals

However, the Riemann integral has its limitations, which are often worked around by invoking the Lebesgue definition of integration.

2.2 Lebesgue Integral

> The Lebesgue integral is another definition of the integral proposed in 1904 by the French mathematician Henri Lebesgue. The formal definition of the Lebesgue integral involves measure theory, which is beyond the scope of the book. A helpful intuitive picture of the Lebesgue integral can be given by thinking of the Lebesgue integral as taking the areas of *horizontal* rectangles instead of the vertical rectangles one encounters in the Riemann definition.

In describing his method, Lebesgue said:[1]

> I have to pay a certain sum, which I have collected in my pocket. I take the bills and coins out of my pocket and give them to the creditor in the order I find them until I have reached the total sum. This is the Riemann integral. But I can proceed differently. After I have taken all the money out of my pocket I order the bills and coins according to identical values and then I pay the several heaps one after the other to the creditor. This is my integral.

In essence, the Lebesgue integral approximates the total area by dividing it into horizontal strips instead of the vertical strips in the Riemann definition. The Lebesgue integral asks "for each y-value, how many x-values produce this value?"

The Lebesgue integral more or less became the "official" integral of research mathematics. However, rarely do engineers and scientists employ this definition. In most scenarios, the Riemann definition suffices. A funny quote is given by the American electrical engineer and computer scientist Richard Hamming who declared confidently

> ...for more than 40 years I have claimed that if whether an airplane would fly or not depended on whether some function that arose in its design was Lebesgue but not Riemann integrable, then I would not fly in it. Would you? Does Nature recognize the difference? I doubt it!

See footnote for the full article[2].

---

[1]Gowers, T., Barrow-Green, J., Leader, I. (2008). *The Princeton companion to mathematics*. Princeton, NJ: Princeton University Press.

[2]Hamming, R. W. (1998). Mathematics on a Distant Planet. *The American Mathematical Monthly*, 105(7), 640. doi: 10.2307/2589247

2.3 The u-substitution

We start by introducing an extremely useful identity.

Proposition. Let $f(x)$ be an integrable function. We then have
$$\int_a^b f(x)\,\mathrm{d}x = \int_a^b f(a+b-x)\,\mathrm{d}x \qquad (2.1)$$

Proof. Let
$$I = \int_a^b f(x)\,\mathrm{d}x$$
Also, let $x = a+b-t$, $\mathrm{d}x = -\mathrm{d}t$. This then transforms I into:

$$I = \int_b^a f(a+b-t)(-\mathrm{d}t)$$
$$= \int_a^b f(a+b-t)\,\mathrm{d}t$$

Rewriting using x,
$$I = \int_a^b f(a+b-x)\,\mathrm{d}x$$

$\square$

This formula is akin to "integrating backwards." It will be used extensively throughout this book, and is especially useful for trigonometric integrals. To start the chapter off, we will begin with an elegant result involving (2.1).

Example 1: Evaluate $\displaystyle\int_0^\pi \frac{x \sin x}{1 + \cos^2 x}\,\mathrm{d}x$

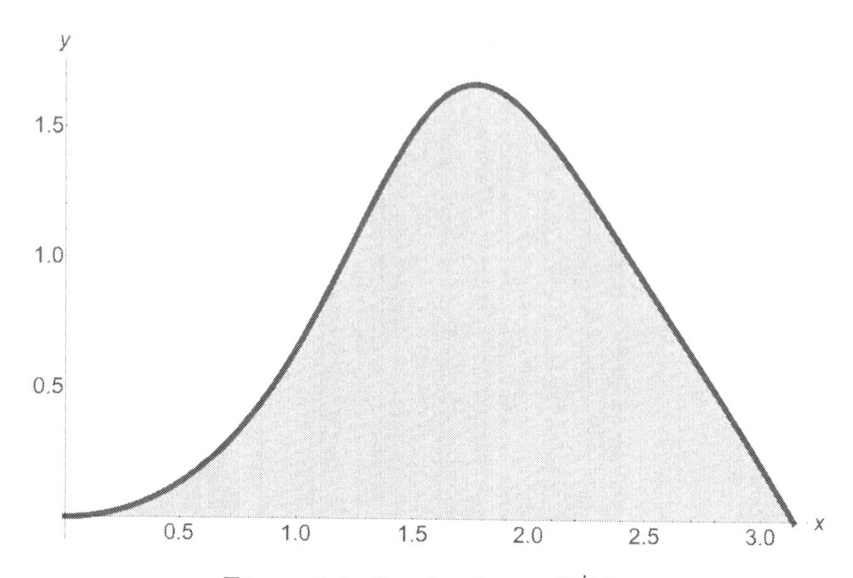

Figure 2.5: Graph of $y = \frac{x \sin x}{1+\cos^2 x}$

Solution

Using (2.1) gives:

$$I = \int_0^\pi \left[\frac{(\pi - x)\sin(\pi - x)}{1 + \cos^2(\pi - x)} \right] dx$$

Adding our original integral with the transformed integral above,

$$2I = \int_0^\pi \left[\frac{x \sin x}{1 + \cos^2 x} + \frac{(\pi - x)\sin(\pi - x)}{1 + \cos^2(\pi - x)} \right] dx$$

$$I = \frac{1}{2} \int_0^\pi \left[\frac{x \sin x}{1 + \cos^2 x} + \frac{(\pi - x)\sin(\pi - x)}{1 + \cos^2(\pi - x)} \right] dx$$

We can simplify by using the identities $\sin(\pi - x) = \sin x$,

2.3. THE U-SUBSTITUTION

$\cos^2(\pi - x) = \cos^2 x$:

$$I = \frac{1}{2} \int_0^\pi \left[\frac{\pi \sin x}{1 + \cos^2 x} \right] dx$$

We now proceed to make the u-substitution $u = \cos x$, $du = -\sin x \, dx$

$$\frac{\pi}{2} \int_1^{-1} \frac{-du}{1 + u^2} = \left[\frac{\pi}{2} \arctan u \right]_{-1}^{1} = \frac{\pi^2}{4}$$

Evaluating our integral with Mathematica,

In[7]:= $\int_0^\pi \frac{x \, \text{Sin}[x]}{1 + \text{Cos}[x]^2} dx$

Out[7]= $\frac{\pi^2}{4}$

Example 2: Evaluate $\int_0^{\frac{\pi}{2}} \ln(\sin x) dx$

Solution
We will use a unique approach to this problem utilizing our reflection identity, (2.1).

$$I = \int_0^{\frac{\pi}{2}} \ln\left(\sin\left(\frac{\pi}{2} - x\right)\right) dx = \int_0^{\frac{\pi}{2}} \ln(\cos x) \, dx$$

Adding our original and transformed integral,

$$2I = \int_0^{\frac{\pi}{2}} \ln(\sin x) \, dx + \int_0^{\frac{\pi}{2}} \ln(\cos x) \, dx$$

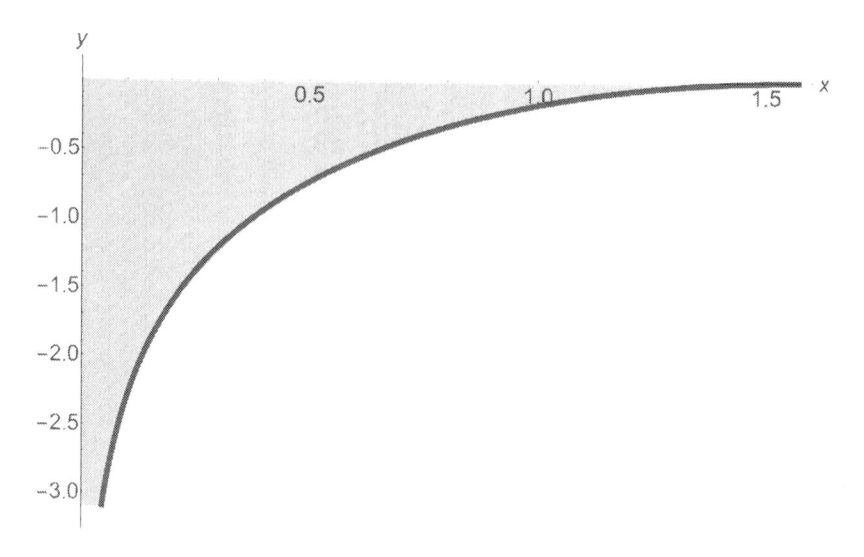

Figure 2.6: Graph of $y = \ln(\sin x)$

$$= \int_0^{\frac{\pi}{2}} \ln(\cos x \sin x) \, dx$$

Using $2 \sin x \cos x = \sin(2x)$, we have:

$$2I = \int_0^{\frac{\pi}{2}} \ln\left(\frac{2 \cos x \sin x}{2}\right) dx = \int_0^{\frac{\pi}{2}} \ln(\sin 2x) - \ln 2 \, dx$$

$$= \int_0^{\frac{\pi}{2}} \ln(\sin 2x) \, dx - \frac{\pi \ln 2}{2}$$

We proceed to make the u-substitution $u = 2x$, $dx = \frac{du}{2}$ which transforms the above expression to:

$$2I = \frac{1}{2} \int_0^{\pi} \ln(\sin u) \, du - \frac{\pi \ln 2}{2}$$

The period of $\sin 2x$ is $\frac{2\pi}{2} = \pi$ from which we can deduce that:

$$\frac{1}{2} \int_0^{\pi} \ln(\sin u) \, du = \int_0^{\frac{\pi}{2}} \ln(\sin u) \, du$$

2.3. THE U-SUBSTITUTION

Therefore,

$$2I = \underbrace{\int_0^{\frac{\pi}{2}} \ln(\sin u)\, du}_{I} - \frac{\pi \ln 2}{2}$$

$$\implies 2I = I - \frac{\pi \ln 2}{2}$$

$$\int_0^{\frac{\pi}{2}} \ln(\sin x)\, dx = -\frac{\pi \ln 2}{2} \tag{2.2}$$

Mathrmatica gives the same answer,

In[8]:= $\int_0^{\frac{\pi}{2}}$ Log[Sin[x]]dx

Out[8]= $-\frac{1}{2} \pi\, \text{Log}[2]$

Example 3: Evaluate $\int_0^{\pi} \frac{x \sin x}{(\cos^2 x + 1)^2} dx$

Solution
Using (2.1),

$$I = \int_0^{\pi} \frac{(\pi - x)\sin(\pi - x)}{(\cos^2(\pi - x) + 1)^2} dx$$

$$= \int_0^{\pi} \frac{\pi \sin(x)}{(\cos^2(x) + 1)^2} dx - \underbrace{\int_0^{\pi} \frac{x \sin(x)}{(\cos^2(x) + 1)^2} dx}_{I}$$

$$\implies I = \frac{\pi}{2} \int_0^{\pi} \frac{\sin(x)}{(\cos^2(x) + 1)^2} dx$$

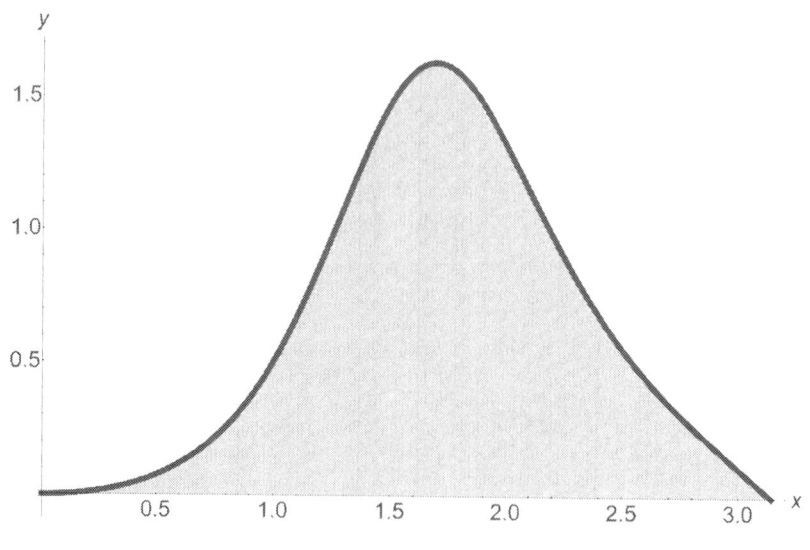

Figure 2.7: Graph of $y = \dfrac{x \sin x}{\left(\cos^2 x + 1\right)^2}$

We apply the intuitive u-substitution $u = \cos x$, $du = -\sin x \, dx$. In doing so we get:

$$I = \frac{\pi}{2} \int_{-1}^{1} \frac{du}{\left(u^2 + 1\right)^2}$$

Substituting again with $u = \tan x$, $du = \sec^2(x) \, dx$ gives

$$I = \frac{\pi}{2} \int_{-\frac{\pi}{4}}^{\frac{\pi}{4}} \frac{\sec^2 x}{(\sec^2 x)^2} dx$$

$$= \frac{\pi}{2} \int_{-\frac{\pi}{4}}^{\frac{\pi}{4}} \cos^2 x \, dx$$

Using

$$\cos^2 x = \frac{1 + \cos 2x}{2}$$

Then gives:

$$I = \frac{\pi}{4} \left(\int_{-\frac{\pi}{4}}^{\frac{\pi}{4}} dx + \int_{-\frac{\pi}{4}}^{\frac{\pi}{4}} \cos(2x) \, dx \right)$$

2.3. THE U-SUBSTITUTION

Which can be easily evaluated to

$$I = \frac{\pi}{4}\left(\frac{\pi}{2} + 1\right)$$

$$I = \frac{\pi^2}{8} + \frac{\pi}{4}$$

Let's see what Mathematica gives,

In[9]:= $\int_0^\pi \frac{x \text{Sin}[x]}{(1+\text{Cos}[x]^2)^2} dx$

Out[9]= $\frac{1}{8} \pi (2+\pi)$

Which, if expanded, equals $\frac{\pi^2}{8} + \frac{\pi}{4}$.

Example 4: Evaluate $\int_0^1 \ln(\Gamma(x)) dx$

Solution

It is worth noting that this result is due to the well-known Swiss polymath Leonhard Euler. We start our solution by using (2.1),

$$I = \int_0^1 \ln(\Gamma(x)) dx = \int_0^1 \ln(\Gamma(1-x)) dx$$

We will now introduce Euler's reflection formula:

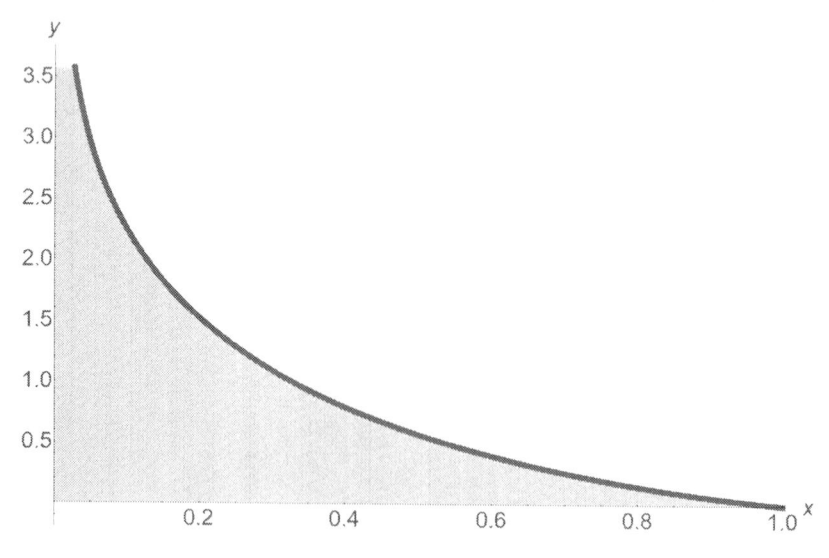

Figure 2.8: Graph of the $y = \ln \Gamma(x)$

> **Theorem**
>
> The infamous Euler's reflection formula is given by:
> $$\Gamma(x)\Gamma(1-x) = \pi \csc \pi x \qquad (2.3)$$
> Where $x \notin \mathbb{Z}$. See (9.3) for proof.

Taking the natural logarithm of both sides in (2.3) gives

$$\ln \Gamma(x) + \ln \Gamma(1-x) = \ln \pi + \ln \csc(\pi x) = \ln \pi - \ln \sin(\pi x)$$

Therefore,

$$2I = \int_0^1 \ln\left(\Gamma(x)\right) + \ln\left(\Gamma(1-x)\right) \, dx = \int_0^1 \ln \pi - \ln \sin(\pi x) \, dx$$

$$2I = \int_0^1 \ln \pi \, dx - \int_0^1 \ln \sin(\pi x) \, dx$$

2.3. THE U-SUBSTITUTION

We proceed to substitute $u = \pi x$, $du = \pi dx$ to get:

$$2I = \ln \pi - \frac{1}{\pi} \int_0^\pi \ln \sin(u) du$$

It is easy to see that $\ln(\sin u)$ is symmetric around $u = \frac{\pi}{2}$. Using this symmetry, we can write

$$\int_0^\pi \ln \sin(u) \, du = 2 \int_0^{\frac{\pi}{2}} \ln \sin(u) \, du$$

We have already evaluated the latter integral in (2.2). Hence,

$$2I = \ln \pi - \frac{2}{\pi} \cdot \frac{-\pi \ln 2}{2}$$
$$= \ln \pi + \ln 2$$

Finally we obtain

$$\int_0^1 \ln(\Gamma(x)) dx = \frac{\ln 2\pi}{2} \qquad (2.4)$$

Evaluating with Mathematica,

In[10]:= $\int_0^1 \text{Log[Gamma[x]] dx}$

Out[10]= $\frac{1}{2} \text{Log}[2\,\pi]$

Example 5: Define $I_1 = \int_0^\infty \frac{1}{1+x^{13}} dx$ and $I_2 = \int_0^\infty \frac{1}{13+x^{13}} dx$. Find $\frac{I_2}{I_1}$.

Solution

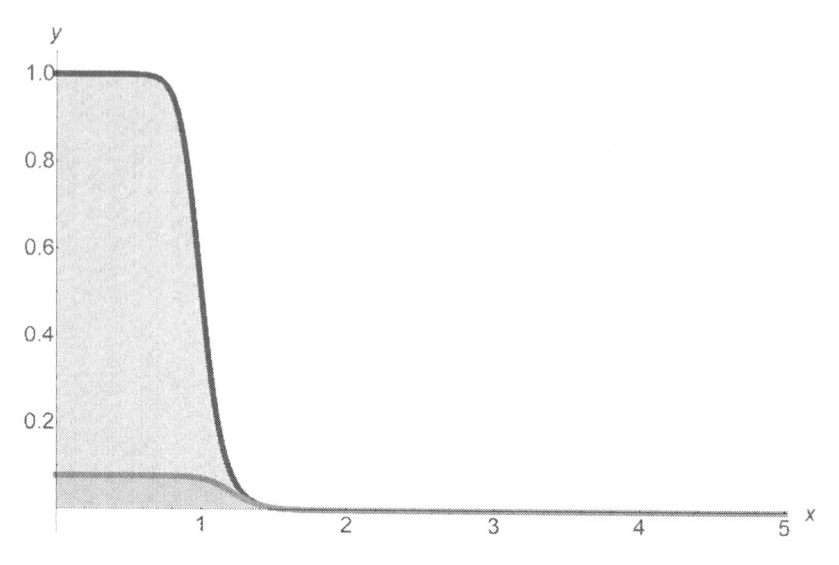

Figure 2.9: Graph of the integrands in I_1 and I_2

Since
$$\int_0^\infty e^{-ax}\,\mathrm{d}x = \frac{1}{a}$$

We can write I_1 as a double integral.

$$I_1 = \int_0^\infty \frac{1}{1+x^{13}}\,\mathrm{d}x = \int_0^\infty \int_0^\infty e^{-y(1+x^{13})}\,\mathrm{d}y\,\mathrm{d}x$$

Let $t = yx^{13}, \mathrm{d}t = 13x^{12}y\,\mathrm{d}x$, which then transforms I_1 into:

$$I_1 = \frac{1}{13}\int_0^\infty e^{-t}t^{-\frac{12}{13}}\,\mathrm{d}t \cdot \int_0^\infty e^{-y}y^{-\frac{1}{13}}\,\mathrm{d}y \qquad (2.5)$$

We now will use the same substitution on I_2. In doing so we obtain:

$$I_2 = \int_0^\infty \int_0^\infty e^{-y(13+x^{13})}\,\mathrm{d}y\,\mathrm{d}x$$
$$= \frac{1}{13}\int_0^\infty e^{-t}t^{-\frac{12}{13}}\,\mathrm{d}t \int_0^\infty e^{-13y}y^{-\frac{1}{13}}\,\mathrm{d}y$$

2.3. THE U-SUBSTITUTION

Notice that the only difference is the coefficient of y in the exponential. To get our desired ratio without directly evaluating these integrals, we will change the coefficient of y in I_2 through the u-substitution $u = 13y$, $du = 13dy$.

$$I_2 = \frac{1}{13 \cdot 13} \int_0^\infty e^{-t} t^{-\frac{12}{13}} dt \int_0^\infty e^{-u} \left(\frac{u}{13}\right)^{-\frac{1}{13}} du$$

$$= \frac{13^{\frac{1}{13}}}{13} \left(\frac{1}{13} \int_0^\infty e^{-t} t^{-\frac{12}{13}} dt \int_0^\infty e^{-u} u^{-\frac{1}{13}} du \right)$$

The expression in parentheses is identical to (2.5). Thus,

$$I_2 = 13^{-\frac{12}{13}} I_1 \implies \frac{I_2}{I_1} = 13^{-\frac{12}{13}}$$

We can see if Mathematica can generate an exact value,

In[11]:=
$$\frac{\int_0^\infty \frac{1}{13+x^{13}} dx}{\int_0^\infty \frac{1}{1+x^{13}} dx}$$

Out[11]= $\dfrac{1}{13^{12/13}}$

Example 6: Evaluate $\displaystyle\int_0^1 \frac{\ln\left(\cos\left(\frac{\pi x}{2}\right)\right)}{x^2 + x} dx$

Solution
We will split the integrand using partial fractions since

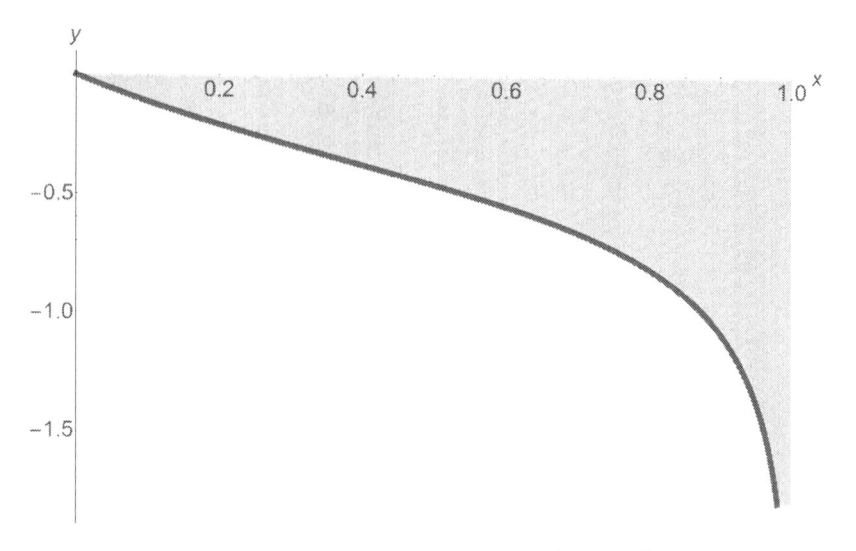

Figure 2.10: Graph of $y = \frac{\ln\left(\cos\left(\frac{\pi x}{2}\right)\right)}{x^2+x}$

$$x^2 + x = (x)(x+1)$$

$$\therefore I = \int_0^1 \frac{\ln \cos\left(\frac{\pi x}{2}\right)}{x} \, dx - \int_0^1 \frac{\ln \cos\left(\frac{\pi x}{2}\right)}{x+1} \, dx$$

Using the trigonometric identity $2\sin x \cos x = \sin 2x$ we can derive that $\cos\left(\frac{\pi x}{2}\right) = \frac{\sin(\pi x)}{2\sin\left(\frac{\pi x}{2}\right)}$. Thus,

$$I = \int_0^1 \frac{\ln\left(\frac{\sin(\pi x)}{2\sin\left(\frac{\pi x}{2}\right)}\right)}{x} \, dx - \int_0^1 \frac{\ln \cos\left(\frac{\pi x}{2}\right)}{x+1} \, dx$$

By the reflection property of integration,

$$I = \int_0^1 \frac{\ln\left(\frac{\sin(\pi x)}{2}\right)}{x} \, dx - \int_0^1 \frac{\ln\left(\sin\left(\frac{\pi x}{2}\right)\right)}{x} \, dx - \int_0^1 \frac{\ln \cos\left(\frac{\pi x}{2}\right)}{x+1} \, dx$$

2.3. THE U-SUBSTITUTION

$$= \underbrace{\int_0^1 \frac{\ln\left(\frac{\sin(\pi x)}{2}\right)}{x}\,dx}_{(1)} - \underbrace{\int_0^1 \frac{\ln\left(\sin\left(\frac{\pi(1-x)}{2}\right)\right)}{1-x}\,dx}_{(2)} - \underbrace{\int_0^1 \frac{\ln\cos\left(\frac{\pi x}{2}\right)}{x+1}\,dx}_{(3)}$$

We will proceed to substitute $u = -x$, $du = -dx$ in (2)

$$I = \underbrace{\int_0^1 \frac{\ln\left(\frac{\sin(\pi x)}{2}\right)}{x}\,dx}_{(1)} - \underbrace{\underbrace{\int_{-1}^0 \frac{\ln\left(\cos\left(\frac{\pi u}{2}\right)\right)}{u+1}\,du}_{(2)} - \underbrace{\int_0^1 \frac{\ln\left(\cos\left(\frac{\pi x}{2}\right)\right)}{x+1}\,dx}_{(3)}}_{\text{Same integrand}}$$

Combining (2) and (3) we obtain

$$I = \underbrace{\int_0^1 \frac{\ln\left(\frac{\sin(\pi x)}{2}\right)}{x}\,dx}_{(1)} - \underbrace{\int_{-1}^1 \frac{\ln\left(\cos\left(\frac{\pi x}{2}\right)\right)}{x+1}\,dx}_{(2^*)}$$

Now, we substitute $u = 1 + x$, $du = dx$ into (2^*)

$$I = \int_0^1 \frac{\ln\left(\frac{\sin(\pi x)}{2}\right)}{x}\,dx - \int_0^2 \frac{\ln\left(\sin\left(\frac{\pi u}{2}\right)\right)}{u}\,du$$

Now, since our integrands are not defined at 0, we need to take the limit as the lower bound approaches 0. Notice that if we were attempting to evaluate the indefinite integral, this step would not be needed at this time. However, it is not possible to derive the above indefinite integrals (anti-derivatives) in standard mathematical functions, so we have to go with the definite integral route.

$$I = \lim_{h \to 0} \left[\int_h^1 \frac{\ln\left(\frac{\sin(\pi x)}{2}\right)}{x}\,dx - \int_h^2 \frac{\ln\left(\sin\left(\frac{\pi x}{2}\right)\right)}{x}\,dx \right]$$

$$= \lim_{h \to 0} \left[\underbrace{\int_h^1 \frac{\ln(\sin(\pi x))}{x} dx}_{(1)} - \underbrace{\int_h^2 \frac{\ln\left(\sin\left(\frac{\pi x}{2}\right)\right)}{x} dx}_{(2^*)} - \int_h^1 \frac{\ln 2}{x} dx \right]$$

Substituting again into (2^*) with $u = \frac{x}{2}$, $du = \frac{dx}{2}$

$$I = \lim_{h \to 0} \left[\underbrace{\int_h^1 \frac{\ln(\sin(\pi x))}{x} dx}_{(1)} - \underbrace{\int_{\frac{h}{2}}^1 \frac{\ln(\sin(\pi u))}{u} du}_{(2^*)} + \ln 2 \ln h \right]$$

Combining (1) and (2^*) we obtain:

$$I = \lim_{h \to 0} \left[-\int_{\frac{h}{2}}^h \frac{\ln(\sin(\pi x))}{x} dx + \ln 2 \ln h \right]$$

As $x \to 0$, $\sin x \approx x$. Since $h \to 0$, we can use this in the above integral to get

$$I = \lim_{h \to 0} \left[-\int_{\frac{h}{2}}^h \frac{\ln(\pi x)}{x} dx + \ln 2 \ln h \right]$$

$$= \lim_{h \to 0} \left[-\int_{\frac{h}{2}}^h \frac{\ln(x)}{x} dx - \int_{\frac{h}{2}}^h \frac{\ln \pi}{x} dx + \ln 2 \ln h \right]$$

Our first integral above can be calculated using

$$\int \frac{\ln x}{x} dx = \frac{(\ln x)^2}{2}$$

Therefore,

$$I = \lim_{h \to 0} \left[-\frac{1}{2} \left[(\ln h)^2 - \left(\ln \frac{h}{2} \right)^2 \right] - \left[\ln \pi \left(\ln h - \ln \frac{h}{2} \right) \right] + \ln 2 \ln h \right]$$

2.3. THE U-SUBSTITUTION

$$= \lim_{h \to 0} \left[-\frac{1}{2} \left(\ln h + \ln \frac{h}{2} \right) \left(\ln h - \ln \frac{h}{2} \right) + \ln 2 \ln h - \ln \pi \ln 2 \right]$$

Where in the last line we used the identity $a^2 - b^2 = (a+b)(a-b)$. Continuing with our solution,

$$I = \lim_{h \to 0} \left[-\frac{1}{2} \left(\ln \frac{h^2}{2} \right) (\ln 2) + \ln 2 \ln h - \ln \pi \ln 2 \right]$$

$$= \lim_{h \to 0} \left[-\frac{1}{2} (2 \ln h - \ln 2)(\ln 2) + \ln 2 \ln h - \ln \pi \ln 2 \right]$$

The terms containing h will cancel out so we are only left with:

$$I = \frac{\ln^2 2}{2} - \ln 2 \ln \pi$$

Oh well, that was a lot of work! To be confident in our result, we can check with Mathematica. Unfortunately, Mathematica only provides a numerical answer. However, we can check the difference between our value above and Mathematica's numerical approximation,

In[12]:= $N[\int_0^1 \frac{\text{Log}[\text{Cos}[\frac{\pi x}{2}]]}{x^2 + x} dx]$
  $-(\frac{\text{Log}[2]^2}{2} - \text{Log}[\pi] \, \text{Log}[2])$

Out[12]= -8.881784197001252*10^-16

Here, $N[\cdot]$ denotes Mathematica's numerical value function.

Example 7: Evaluate $\int_0^1 \dfrac{x \ln x}{\sqrt{1-x^2}}\,dx$

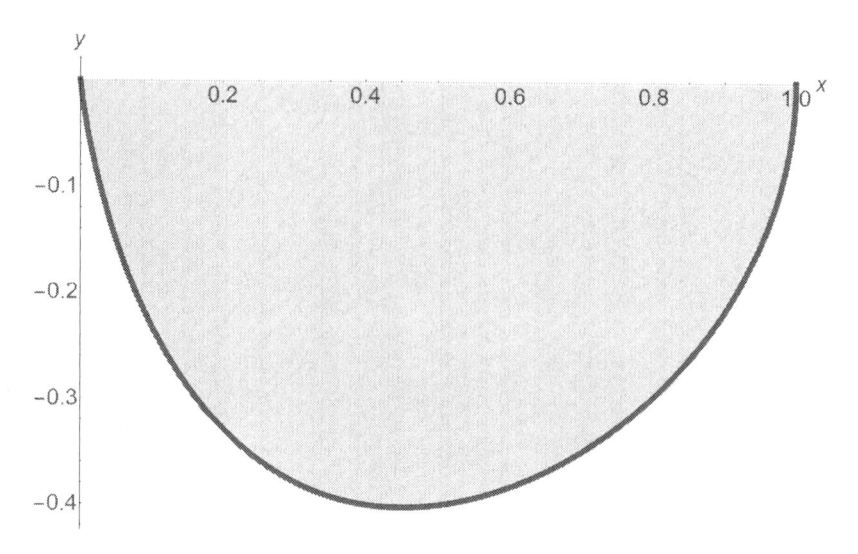

Figure 2.11: Graph of $y = \dfrac{x \ln x}{\sqrt{1-x^2}}$

Solution
Let $u = \sqrt{1-x^2}$, $du = -\dfrac{x}{\sqrt{1-x^2}}dx$. Our integral is then

$$I = -\int_1^0 \ln\sqrt{1-u^2}\,du$$

$$= \frac{1}{2}\int_0^1 \ln\left(1-u^2\right)\,du$$

We can split I into two integrals using basic logarithm rules,

$$I = \frac{1}{2}\int_0^1 \ln(1+u) + \ln(1-u)\,du$$

$$= \frac{1}{2}\left[\int_0^1 \ln(1+u)\,du + \int_0^1 \ln(1-u)\,du\right]$$

$$= \frac{1}{2}\left[\Big[u\ln u - u\Big]_1^2 - \Big[u\ln u - u\Big]_1^0\right]$$

Since $\lim_{u\to 0} u\ln u = 0$,

$$\therefore \int_0^1 \frac{x\ln x}{\sqrt{1-x^2}}\,dx = \frac{2\ln 2 - 2}{2} = \ln 2 - 1 \qquad (2.6)$$

Checking with Mathematica,

In[13]:= $\int_0^1 \frac{x\text{Log}[x]}{\sqrt{1-x^2}}dx$

Out[13]= -1+Log[2]

We saw how u-substitutions can make seemingly hard integrals simple. In the next section, we will see some other elementary ways to solve integrals.

2.4 Other Problems

> **Theorem**
>
> Integration by parts, also known as IBP, is one of the most common integration techniques. It is derived from the product rule,
>
> $$(uv)' = u'v + v'u$$
>
> And is its analogue in integration. The rule states that
>
> $$\int u\,dv = uv - \int v\,du \qquad (2.7)$$
>
> For two continuous and differentiable functions u and v.

For definite integrals, we have

$$\int_a^b u\,dv = [uv]_a^b - \int_a^b v\,du$$

This result can be extended to three functions, u, v, w:

$$\int_a^b uv\,dw = \left[uvw\right]_a^b - \int_a^b uw\,dv - \int_a^b vw\,du$$

This technique was first discovered by the English mathematician Brook Taylor, who also discovered Taylor series[3].

Proof. Consider the statement of the product rule:

$$(uv)' = u'v + v'u$$

Define $u = u(x)$ and $v = v(x)$, i.e. u, v are functions of x. Integrating both sides with respect to x,

$$\int (uv)' = \int u'v + v'u$$

$$uv = \int u'v + \int v'u$$

Rearranging this formula gives:

$$\int v'u = uv - \int u'v$$

---

[3]TAYLOR, B. (1715). *Methodus Incrementorum directa et inversa.* Londini: Apud Gul. Innys.

2.4. OTHER PROBLEMS

Expressing this in Leibniz notation,

$$\int u \, dv = uv - \int v \, du$$

This can be easily applied to definite integrals

$$\int_a^b u \, dv = [uv]_a^b - \int_a^b v \, du$$

By the first fundamental theorem of calculus. □

Now, let's apply our skills!

Example 8: Find $\int_0^\infty \arcsin e^{-x} \, dx$

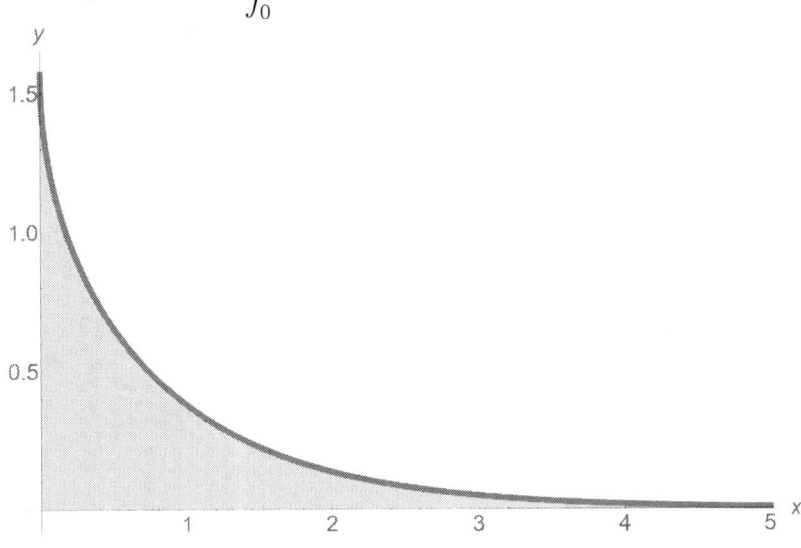

Figure 2.12: Graph of $y = \arcsin e^{-x}$

Solution

Sometimes it is more convenient to do a substitution in terms of functions on both sides. In essence, the substitution form is:

$$f(u) = g(x)$$

Instead of $u = f^{-1}(g(x))$. Also, the introduction of another variable, namely u, is not necessary. One can do a direct substitution in the same variable. In this case, we would like to substitute $e^{-x} \to \sin x$. For the sake of ease, we can also write this as $e^{-x} = \sin u$, $e^{-x} dx = -\cos u \, du \implies dx = -\cot u \, du$ and obtain that we need to substitute $dx \to -\cot x \, dx$. As one gains more practice with such procedures, it will become almost second nature!

$$I = \int_0^{\frac{\pi}{2}} x \cot x \, dx$$

Using integration by parts with $u = x$, $du = dx$ and $dv = \cot x \, dx$, $v = \ln \sin x$ leads us to

$$I = [x \ln \sin x]_0^{\frac{\pi}{2}} - \int_0^{\frac{\pi}{2}} \ln \sin x \, dx$$

$$I = \frac{\pi}{2} \ln 2$$

See (2.2) for the evaluation of the integral. We can check Mathematica to verify our solution,

In[14]:= $\int_0^\infty \text{ArcSin}[e^{-x}] \, dx$

Out[14]= $\frac{1}{2} \pi \, \text{Log}[2]$

Example 9: Find $\int \sin(2019x) \sin^{2017} x \, dx$

2.4. OTHER PROBLEMS

Solution
We will show that for any n,

$$I_n = \int \sin(nx) \sin^{n-2} x \, dx = \frac{\sin^{n-1} x \sin((n-1)x)}{n-1} + C$$

We will start by splitting up n into $n-1$ and 1

$$I_n = \int \sin((n-1)x + x) \sin^{n-2} x \, dx$$

By using the fact that

$$\sin(\alpha + \beta) = \sin \alpha \cos \beta + \sin \beta \cos \alpha$$

We obtain:

$$I_n = \int \sin((n-1)x) \cos(x) \sin^{n-2} x + \sin x \cos((n-1)x) \sin^{n-2} x \, dx$$

$$= \int \sin((n-1)x) \cos(x) \sin^{n-2} x + \cos((n-1)x) \sin^{n-1} x \, dx$$

We proceed to multiply by $\frac{n-1}{n-1}$ to get

$$I_n = \frac{1}{n-1} \int \sin((n-1)x) \left((n-1) \cos(x) \sin^{n-2} x\right)$$
$$+ (n-1) \cos((n-1)x) \sin^{n-1} x \, dx$$

Notice that

$$\left(\sin^{n-1} x\right)' = (n-1) \cos(x) \sin^{n-2} x$$

And

$$\left(\sin((n-1)x)\right)' = (n-1) \cos((n-1)x)$$

Therefore,

$$I_n = \frac{1}{n-1} \int \sin((n-1)x)(\sin^{n-1} x)' + \left(\sin((n-1)x)\right)' \sin^{n-1} x \, dx$$

The integrand is simply $\left(\sin^{n-1} x \sin((n-1)x)\right)'$ by the product rule. Therefore,

$$I_n = \frac{1}{n-1} \int \left(\sin((n-1)x) \sin^{n-1} x\right)' \, dx$$
$$= \frac{\sin((n-1)x) \sin^{n-1} x}{n-1} + C$$

We can now plug in our desired value, $n = 2019$

$$I_{2019} = \frac{\sin(2018x) \sin^{2018} x}{2018} + C$$

Mathematica gives

In[15]:= $\int \text{Sin}[2019 \, x] \, \text{Sin}[x]^{2017} \, dx$

Out[15]= $\dfrac{\text{Sin}[x]^{2018} \, \text{Sin}[2018 \, x]}{2018}$

Example 10: Find $\displaystyle\lim_{n \to \infty} \int_0^{\frac{\pi}{2}} \sqrt[n]{\sin^n x + \cos^n x} \, dx$

2.4. OTHER PROBLEMS

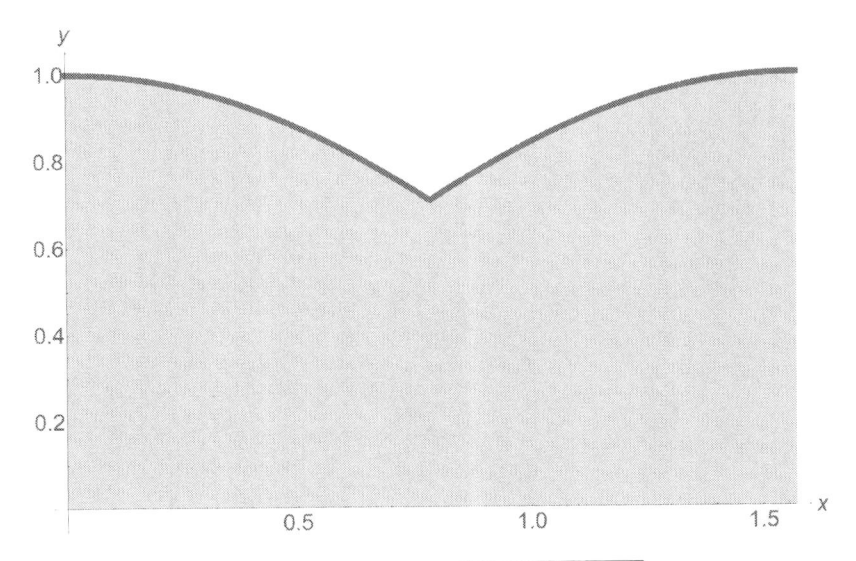

Figure 2.13: Graph of the $y = \sqrt[n]{\sin^n x + \cos^n x}$ for $n = 1000$

Solution
Let

$$A = \lim_{n \to \infty} \sqrt[n]{\sin^n x + \cos^n x} = \lim_{n \to \infty} \left(\cos x \sqrt[n]{1 + \tan^n x} \right)$$

Where we factored out $\cos^n x$ in the last step. Also, let

$$B = \lim_{n \to \infty} \sqrt[n]{1 + \tan^n x}$$

$$\implies \ln B = \lim_{n \to \infty} \frac{\ln(1 + \tan^n x)}{n}$$

Notice that when $x \in \left[0, \frac{\pi}{4}\right), \tan x \in [0, 1)$. Hence, when $x \in \left[0, \frac{\pi}{4}\right)$, we know that $\ln B = 0$ and $B = 1$ since as $n \to \infty$, $\tan^n x = 0$. On the other hand, when $x \in \left[\frac{\pi}{4}, \frac{\pi}{2}\right), \tan x \in [1, \infty)$. We can then deduce that $B = \sqrt[n]{\tan^n x} = \tan x$. Our integrand is then simply

$$B\cos x = \begin{cases} \cos x & , x \in \left[0, \frac{\pi}{4}\right) \\ \cos x \tan x = \sin x & , x \in \left[\frac{\pi}{4}, \frac{\pi}{2}\right) \end{cases}$$

Even though $\tan x$ diverges at $x = \frac{\pi}{2}$, we can see that

$$\lim_{x \to \frac{\pi}{2}} A = \lim_{x \to \frac{\pi}{2}} B \cos x = 1$$

Therefore,

$$\lim_{n \to \infty} \int_0^{\frac{\pi}{2}} \sqrt[n]{\sin^n x + \cos^n x}\, dx$$

$$= \int_0^{\frac{\pi}{4}} \cos x\, dx + \int_{\frac{\pi}{4}}^{\frac{\pi}{2}} \sin x\, dx$$

$$= [\sin x]_0^{\frac{\pi}{4}} - [\cos x]_{\frac{\pi}{4}}^{\frac{\pi}{2}}$$

$$= \sqrt{2}$$

For $n = 1000$, Mathematica gives ≈ 1.414.

Example 11: Find an expression for $\int_0^\infty \frac{1}{1+x^n}\, dx$ for $n > 1$.

Solution
We will express this integral as a double integral similar to what we did in (2.5). In doing so we obtain:

$$I = \int_0^\infty \int_0^\infty e^{-y(1+x^n)}\, dx\, dy$$

2.4. OTHER PROBLEMS

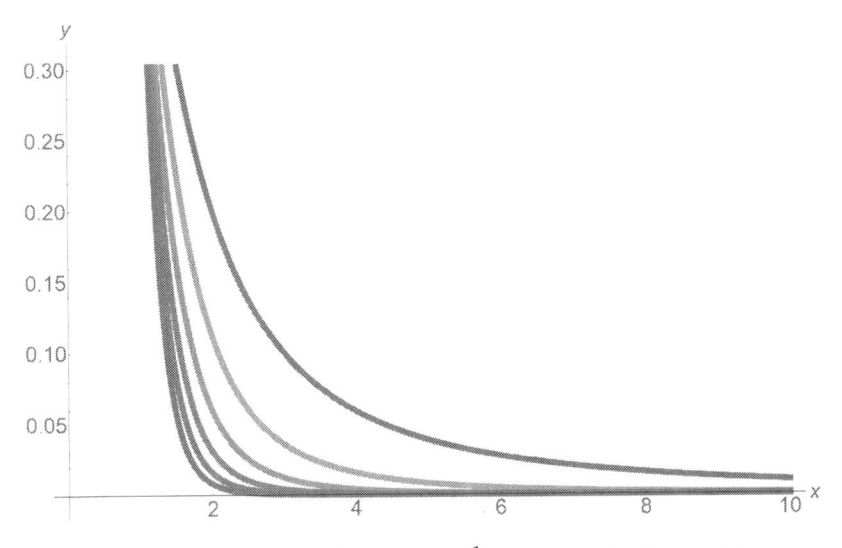

Figure 2.14: Graph of $y = \frac{1}{1+x^n}$ for $n = 2, 3, \cdots, 7$

Then we substitute $u = yx^n$, $du = nx^{n-1}y\,dx$. Therefore, our double integral expressed as a product of two integrals is

$$I = \int_0^\infty \int_0^\infty e^{-y}e^{-yx^n}\,dx\,dy = \frac{1}{n}\int_0^\infty e^{-y}y^{-\frac{1}{n}}\,dy \int_0^\infty u^{\frac{1}{n}-1}e^{-u}\,du$$

Notice that the first integral is simply $\Gamma\left(1 - \frac{1}{n}\right)$ and the second integral is $\Gamma\left(\frac{1}{n}\right)$. Here, Γ denotes the Gamma function as usual (See (1.7)). Using Euler's reflection formula, (2.3), we can write this as

$$\int_0^\infty \frac{1}{1+x^n}\,dx = \frac{\pi}{n}\csc\left(\frac{\pi}{n}\right) \tag{2.8}$$

Checking with Mathematica,

In[16]:= $\int_0^\infty \frac{1}{1+x^n}\,dx$

Out[16]= ConditionalExpression$\left[\dfrac{\pi\ \text{Csc}[\frac{\pi}{n}]}{n}, \text{Re}[n] > 1\right]$

Example 12: Evaluate the integral $\int_0^\infty \dfrac{\ln x^2}{x^2 + a^2}\,dx$ where a is a constant.

Solution

We begin by substituting $x = \dfrac{a^2}{u}$, $dx = -\dfrac{a^2}{u^2}\,du$

$$I(a) = 2\int_0^\infty \dfrac{a^2(2\ln a - \ln u)}{u^2\left(\dfrac{a^4}{u^2} + a^2\right)}\,du$$

$$= 2\int_0^\infty \dfrac{2\ln a - \ln u}{a^2 + u^2}\,du$$

$$= 4\ln a \int_0^\infty \dfrac{du}{a^2 + u^2} - 2\underbrace{\int_0^\infty \dfrac{\ln u}{a^2 + u^2}\,du}_{I(a)}$$

Therefore,

$$2I(a) = 4\ln a \int_0^\infty \dfrac{du}{a^2 + u^2}$$

The substitution $u = a\tan y$, $du = a\sec^2 y\,dy$ gives:

$$I(a) = 2\ln a \int_0^{\frac{\pi}{2}} \dfrac{a\sec^2 y}{a^2(\tan^2 y + 1)}\,dy = \dfrac{2\ln a}{a}\int_0^{\frac{\pi}{2}} dy$$

$$\Longrightarrow I(a) = \dfrac{\pi \ln a}{a}$$

Here, Mathematica returns the same result.

In[17]:= $\int_0^\infty \dfrac{\text{Log}[x^2]}{x^2 + a^2}\,dx$

2.4. OTHER PROBLEMS

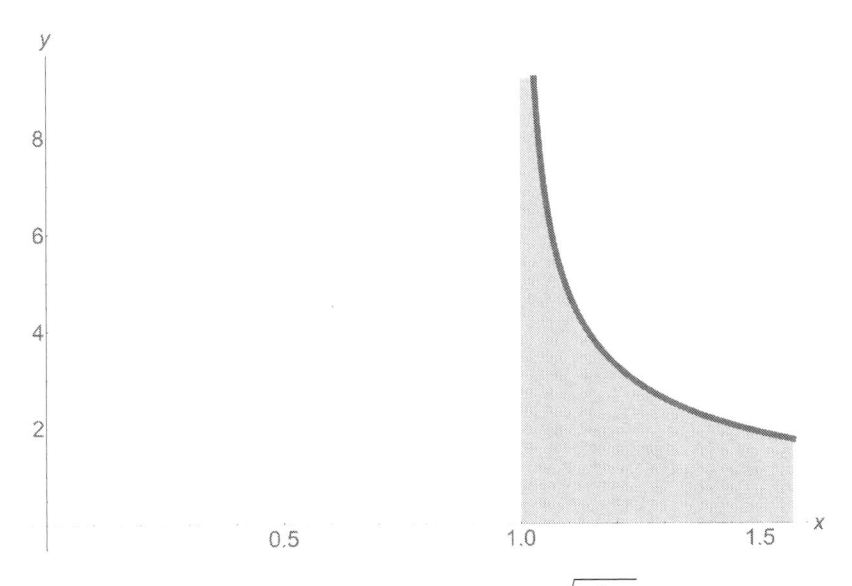

Figure 2.15: Graph of $y = \dfrac{\sqrt{\frac{4x^2+1}{x^2-1}}}{x}$

Example 13: Find $\displaystyle\int_1^{\sqrt{6}} \dfrac{\sqrt{\frac{4x^2+1}{x^2-1}}}{x}\,\mathrm{d}x$

Solution
Let $u = \sqrt{\dfrac{4x^2+1}{x^2-1}}$. Therefore,

$$u^2 = \frac{4x^2+1}{x^2-1}$$

$$= \frac{4(x^2-1)+5}{x^2-1}$$

$$\implies u^2 - 4 = \frac{5}{x^2-1}$$

We now need to obtain an expression for $\mathrm{d}x$ in terms of u only. With some algebraic manipulation we obtain:

$$\frac{1}{u^2-4} = \frac{x^2-1}{5}$$

$$\frac{5}{u^2-4} + \frac{u^2-4}{u^2-4} = x^2$$

$$x^2 = \frac{u^2+1}{u^2-4} \qquad (2.9)$$

Differentiating both sides,

$$2x\,dx = \frac{-10u}{(u^2-4)^2}du \qquad (2.10)$$

Now we proceed to find $\frac{dx}{x}$. We start by writing

$$\frac{dx}{x} = \frac{2x}{2x^2}dx$$

Plugging in our results from (2.9) and (2.10),

$$\frac{dx}{x} = \frac{-10u}{(u^2-4)^2}du \cdot \frac{u^2-4}{2(u^2+1)}$$

$$= \frac{-5u}{(u^2-4)(u^2+1)}du$$

Therefore,

$$I = \int_{\sqrt{5}}^{\infty} \frac{5u^2}{(u^2-4)(u^2+1)}du$$

Using the method of partial fractions we obtain:

$$I = \int_{\sqrt{5}}^{\infty} \frac{4}{u^2-4} + \frac{1}{u^2+1}du$$

$$= \int_{\sqrt{5}}^{\infty} \left(\frac{1}{u-2} - \frac{1}{u+2}\right)du + \Big[\arctan u\Big]_{\sqrt{5}}^{\infty}$$

2.4. OTHER PROBLEMS

$$= \left[\ln\left(\frac{u-2}{u+2}\right)\right]_{\sqrt{5}}^{\infty} + \frac{\pi}{2} - \arctan\sqrt{5}$$

$$= -\ln\left(\frac{\sqrt{5}-2}{\sqrt{5}+2}\right) + \frac{\pi}{2} - \arctan\sqrt{5}$$

Simplifying,

$$I = 2\ln\left(2+\sqrt{5}\right) + \frac{\pi}{2} - \arctan\sqrt{5}$$

Let's see what Mathematica gives,

In[18]:= $\int_1^{\sqrt{6}} \frac{\sqrt{\frac{4x^2+1}{x^2-1}}}{x}\,dx$

Out[18]= $2\,\text{ArcSinh}[2] + \text{ArcTan}[\frac{1}{\sqrt{5}}]$

Which are equivalent since

$$\sinh^{-1} x = \ln\left(\sqrt{x^2+1}+x\right)$$

And

$$\arctan x = \frac{\pi}{2} - \arctan x^{-1}$$

Example 14: Evaluate $\int_0^1 \frac{\arcsin\left(\frac{2x}{1+x^2}\right)}{1+x^2}\,dx$

Solution

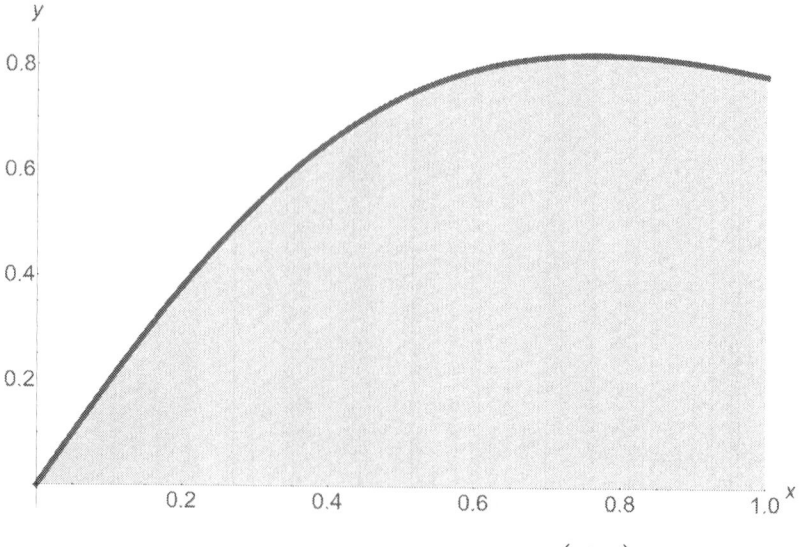

Figure 2.16: Graph of $y = \dfrac{\arcsin\left(\frac{2x}{1+x^2}\right)}{1+x^2}$

In general, a $1 + x^2$ term in the denominator is a tell-tale sign that the substitution $x = \tan u$ will work, so we begin by letting $x = \tan u$, $dx = \sec^2 u \, du$. Our integral is then transformed to

$$I = \int_0^{\frac{\pi}{4}} \frac{\arcsin\left(\frac{2\tan u}{1+\tan^2 u}\right)}{1+\tan^2 u} \cdot \sec^2 u \, du$$

$$I = \int_0^{\frac{\pi}{4}} \arcsin\left(\frac{2\tan u}{1+\tan^2 u}\right) du$$

We proceed to apply the trigonometric identity:

$$\sin u = \frac{2\tan \frac{u}{2}}{1+\tan^2 \frac{u}{2}} \implies \sin 2u = \frac{2\tan u}{1+\tan^2 u}$$

Which gives

2.4. OTHER PROBLEMS

$$I = \int_0^{\frac{\pi}{4}} \arcsin(\sin 2u) \, du$$

On the interval $[0, \frac{\pi}{4}]$, $\arcsin(\sin 2u) = 2u$. Therefore,

$$I = \int_0^{\frac{\pi}{4}} 2u \, du$$

$$I = \frac{\pi^2}{16} \tag{2.11}$$

Mathematica also gives this value as well

In[19]:= $\int_0^1 \frac{\text{ArcSin}[\frac{2x}{1+x^2}]}{1+x^2} dx$

Out[19]= $\frac{\pi^2}{16}$

We can end this chapter with an interesting formula

Theorem

Let $f(x)$ be an even Riemann-integrable function on $[-\alpha, \alpha]$ and $g(x)$ an odd Riemann-integrable function on $[-\alpha, \alpha]$, where $\alpha \in \mathbb{R}$. We then have:

$$\int_{-\alpha}^{\alpha} \frac{f(x)}{1 + b^{g(x)}} dx = \int_0^{\alpha} f(x) dx \tag{2.12}$$

For any $b \in \mathbb{R}^+$.

Proof. Notice that we can write (2.12) as:

$$\int_{-\alpha}^{\alpha} \frac{f(x)}{1+b^{g(x)}}dx = \underbrace{\int_{-\alpha}^{0} \frac{f(x)}{1+b^{g(x)}}dx}_{I_1} + \underbrace{\int_{0}^{\alpha} \frac{f(x)}{1+b^{g(x)}}dx}_{I_2}$$

The substitution $u = -x$, $du = -dx$ in I_1 gives:

$$I_1 = \int_0^{\alpha} \frac{f(-u)}{1+b^{g(-u)}}du$$

Using the fact that $f(u)$ is even and $g(u)$ is odd, we have

$$I_1 = \int_0^{\alpha} \frac{f(u)}{1+b^{-g(u)}}du$$

Multiplying by $\frac{b^{g(u)}}{b^{g(u)}}$,

$$I_1 = \int_0^{\alpha} \frac{f(u)b^{g(u)}}{1+b^{g(u)}}du$$

Hence,

$$\int_{-\alpha}^{\alpha} \frac{f(x)}{1+b^{g(x)}}dx = I_1 + I_2$$

$$= \int_0^{\alpha} \frac{f(u)b^{g(u)}}{1+b^{g(u)}}du + \int_0^{\alpha} \frac{f(x)}{1+b^{g(x)}}dx$$

$$= \int_0^{\alpha} \frac{\left(1+b^{g(u)}\right)f(u)}{1+b^{g(u)}}du$$

$$= \int_0^{\alpha} f(u)\,du$$

2.4. OTHER PROBLEMS

$$= \int_0^\alpha f(x)\,dx$$

□

Example 15: Evaluate $\int_{-\pi}^{\pi} \dfrac{\sin^2 x}{1+e^{x^3}}\,dx$

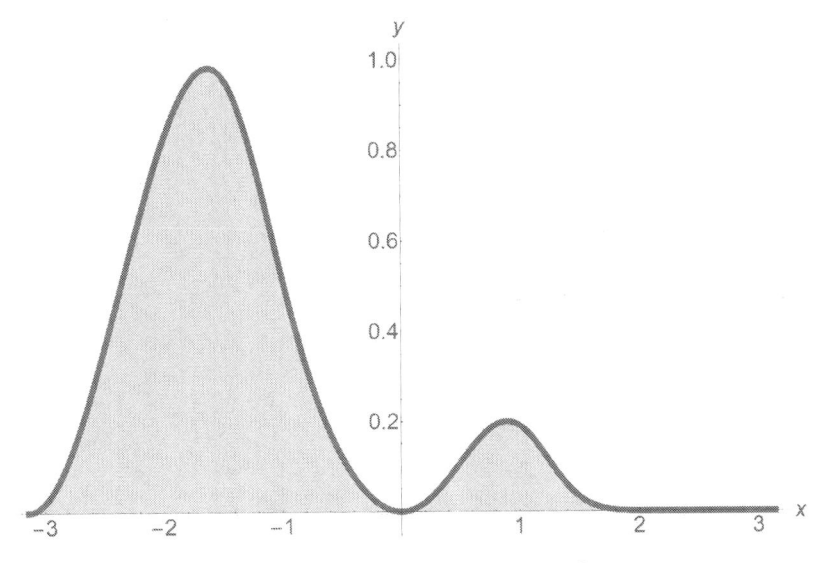

Figure 2.17: Graph of $y = \frac{\sin^2 x}{1+e^{x^3}}$

Solution

Notice that we can use (2.12) since $f(x) = \sin^2 x$ is an even function on $[-\pi, \pi]$ and $g(x) = x^3$ is an odd function on $[-\pi, \pi]$. We then have:

$$I = \int_{-\pi}^{\pi} \frac{\sin^2 x}{1+e^{x^3}}\,dx$$

$$= \int_0^{\pi} \sin^2 x\,dx$$

$$= \frac{1}{2}\Big[x - \sin x \cos x\Big]_0^\pi$$

$$= \frac{\pi}{2}$$

Verifying with Mathematica,

In[20]:= $\int_{-\pi}^{\pi} \frac{\text{Sin}[x]^2}{1+e^{x^3}} dx$

Out[20]= $\frac{\pi}{2}$

And here we see the beauty of generalization!

In the first two chapters we saw how our elementary calculus tools can solve a plethora of problems through creative application of these tools. In nonstandard mathematical problems such as the ones presented in this book, expanding one's mathematical toolkit is of tremendous help. However, the mastery of one's toolkit will lead to far more ingenious and creative solutions! In the proceeding chapters, we will explore and master additional tools to add to your toolkit.

2.5 Exercise Problems

1) Evaluate $\int_0^\infty \frac{dx}{(1+x)^3 + 1}$ (Hint: $u^3 + 1 = (u^2 - u + 1)(u+1)$)

2.5. EXERCISE PROBLEMS

2) Evaluate $\int_0^1 \frac{\ln(1+x)}{1+x^2} dx$

3) Find $\int_0^{\frac{\pi}{2}} \frac{\sqrt{\tan x}}{\sin x (\sin x + \cos x)} dx$

4) Find $\int_1^{\sqrt{3}} \frac{dx}{(1+x^2)^{3/2}}$

5) Evaluate $\int_0^{\pi} \sqrt{1 - \sin x} \, dx$

6) Evaluate $\int_0^{\pi/2} \frac{dx}{1 + \tan^{2019} x}$ (Hint: 2019 is a distractor)

7) Find $\int_0^{\infty} \frac{\ln x}{x^2 + 2x + 4} dx$ (Hint: Let $u = \frac{4}{x}$)

8) Find $\int_0^{2\pi} \frac{3 - \cos x}{3 + \cos x} dx$ (Hint: Use the substitution $u = \tan \frac{x}{2}$. See (2.11))

9) Show that $\frac{22}{7} > \pi$ using the integral $\int_0^1 \frac{x^4(1-x)^4}{1+x^2} dx$ (Source: 1968 Putnam competition)

10) Define $I(a) = \int_0^{\frac{\pi}{4}} e^x \tan^a x \, dx$. Find $\lim_{a \to \infty} aI(a)$

Chapter 3

Feynman's Trick

3.1 Introduction

A straightforward but extremely effective technique, differentiation under the integral sign is simply a clever use of Leibniz's integral rule. In mathematical pop culture, it is often labeled as "Feynman's trick" after the late American physicist Richard Feynman (1918-1988). Feynman said this in discussing his "trick":

> One thing I never did learn was contour integration. I had learned to do integrals by various methods shown in a book that my high school physics teacher Mr. Bader had given me. The book also showed how to differentiate parameters under the integral sign - It's a certain operation. It turns out that's not taught very much in the universities; they don't emphasize it. But I caught on how to use that method, and I used that one damn tool again and again. So because I was self-taught using that book, I had peculiar methods of doing integrals. The result was that, when guys at MIT or Princeton had trouble doing a certain integral, it was because they couldn't do it with the standard methods they had learned in school. If it was contour integration, they would have found it; if it was a simple series expansion, they would have found it. Then I come along and try differentiating under the integral sign, and often it worked. So I got a great reputation for doing integrals, only because my box of tools was different from everybody else's, and they had tried all their tools on it before giving the problem to me.

The last sentence perhaps best summarizes not only the purpose of this chapter but the entire book as well. Often in engineering and the mathematical sciences, peculiar and nonstandard integrals arise. And unfortunately, few resources are avail-

3.2. DIRECT APPROACH

able for solving such integrals! The goal of this book is to provide the reader with a plethora of techniques and methods to go about such problems. So, with that in mind, let us get started!

> **Theorem**
>
> Let $f(x, \alpha)$ be a differentiable function in α with $\frac{\partial}{\partial \alpha} f$ continuous. Then the following equality holds:
>
> $$\frac{\mathrm{d}}{\mathrm{d}\alpha} \int_a^b f(x, \alpha) \, \mathrm{d}x = \int_a^b \frac{\partial}{\partial \alpha} f(x, \alpha) \, \mathrm{d}x$$

See footnote for a rigorous justification of the above theorem[1].

3.2 Direct Approach

In this method, an integral $f(\alpha)$ is defined with the parameter α such that when the integral is differentiated, a standard integral is found. After deriving a closed form for $f'(\alpha)$, one integrates the closed form and obtains an expression for $f(\alpha)$, of course with a $+C$ at the end.

After an expression is derived for $f(\alpha)$, an initial is needed to determine the added constant "C". In most integrals, $\alpha = 0$ or $\alpha = \infty$ is easy or even trivial to compute and therefore chosen as an initial condition.

Now, what does this look like? Let us take a look at a well-known example.

---

[1] Hijab, O. *Introduction to Calculus and Classical Analysis*. New York: Springer-Verlag, p. 189, 1997.

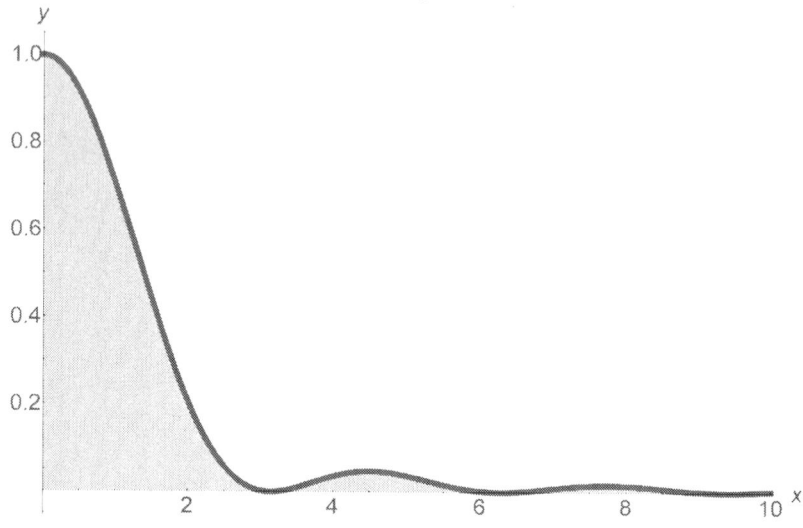

Example 1: Find the value of $\int_0^\infty \frac{\sin^2 x}{x^2} dx$

Figure 3.1: Graph of $y = \frac{\sin^2 x}{x^2}$

Solution

We first use IBP with $u = \sin^2 x$ and $dv = \frac{dx}{x^2}$.

$$I = \int_0^\infty \frac{\sin^2 x}{x^2} dx = \left[-\frac{\sin^2 x}{x} \right]_0^\infty + \int_0^\infty \frac{2 \sin x \cos x}{x} dx$$

$$= 0 + \int_0^\infty \frac{\sin 2x}{x} dx$$

Substituting $2x \to x$,

$$I = \int_0^\infty \frac{\sin x}{x} dx$$

This is the famous **Dirichlet integral**, named after the German mathematician Peter Dirichlet. Now, consider an integral

3.2. DIRECT APPROACH

of the form
$$f(\alpha) = \int_0^\infty \frac{e^{-\alpha x} \sin x}{x} dx$$

Where $\alpha \geq 0$. Our original integral is $I = f(0)$. Differentiating under the integral sign,

$$f'(\alpha) = -\int_0^\infty e^{-\alpha x} \sin x \, dx$$

$$= -\frac{1}{1+\alpha^2}$$

Where we used IBP to get our last result. We can now integrate to get an expression for $f(\alpha)$,

$$f(\alpha) = -\int \frac{1}{1+\alpha^2} d\alpha = -\arctan \alpha + C$$

We know that
$$\lim_{\alpha \to \infty} f(\alpha) = 0 \implies C = \frac{\pi}{2}$$

Hence
$$f(\alpha) = \frac{\pi}{2} - \arctan \alpha$$

Plugging in our desired value,
$$f(0) = \frac{\pi}{2}$$

This example serves as a perfect example of thinking outside the box!

Example 2: Evaluate $\int_0^\infty \frac{x \ln\left(\frac{x^3+1}{x^3}\right)}{1+x^3} dx$

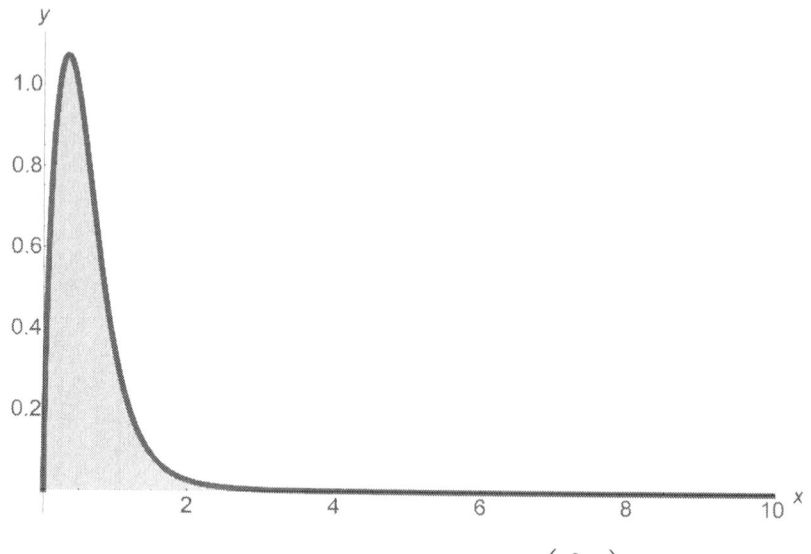

Figure 3.2: Graph of $y = \dfrac{x \ln\left(\frac{x^3+1}{x^3}\right)}{1+x^3}$

Solution

Let

$$f(\alpha) = \int_0^\infty \frac{x \ln\left(\frac{x^3+\alpha}{x^3}\right)}{1+x^3} \, dx$$

$$= \int_0^\infty \frac{x \left(\ln\left(x^3+\alpha\right) - \ln x^3\right)}{1+x^3} \, dx$$

Differentiating under the integral sign,

$$f'(\alpha) = \int_0^\infty \frac{x}{(x^3+\alpha)(1+x^3)} \, dx$$

The substitution $u = x^3$, $dx = \dfrac{du}{3u^{\frac{2}{3}}}$ gives

3.2. DIRECT APPROACH

$$f'(\alpha) = \frac{1}{3}\int_0^\infty \frac{du}{u^{\frac{1}{3}}(\alpha+u)(1+u)}$$

Using the method of partial fractions,

$$f'(\alpha) = \frac{1}{3\alpha - 3}\left[\int_0^\infty \frac{du}{u^{\frac{1}{3}}(1+u)} - \int_0^\infty \frac{du}{u^{\frac{1}{3}}(\alpha+u)}\right] \quad (3.1)$$

We can use the fact that

$$\int_0^\infty \frac{du}{u^{\frac{1}{3}}(u+\alpha)} = \frac{2\pi}{\alpha^{\frac{1}{3}}\sqrt{3}}$$

The proof is left as an exercise to the reader (Hint: Make the substitution $y = u^{\frac{1}{3}}$ and then use the method of partial fractions). Plugging in that result into (3.1) gives:

$$f'(\alpha) = \frac{1}{3\alpha - 3}\left[\frac{2\pi}{\sqrt{3}} - \frac{2\pi}{\alpha^{\frac{1}{3}}\sqrt{3}}\right]$$

Simplifying algebraically,

$$f'(\alpha) = \frac{2\pi}{3\sqrt{3}}\left(\frac{1}{\alpha^{\frac{1}{3}} + \alpha^{\frac{2}{3}} + \alpha}\right) \quad (3.2)$$

Since $f(0) = 0$, we can take the definite integral of (3.2) from 0 to 1 to get our desired integral, $f(1)$.

$$f(1) = \int_0^1 f'(\alpha)\,d\alpha$$

$$= \frac{2\pi}{3\sqrt{3}}\int_0^1 \frac{d\alpha}{\alpha^{\frac{1}{3}} + \alpha^{\frac{2}{3}} + \alpha}$$

$$= \frac{2\pi}{3\sqrt{3}} \int_0^1 \frac{d\alpha}{\alpha^{\frac{1}{3}}\left(\alpha^{\frac{2}{3}} + \alpha^{\frac{1}{3}} + 1\right)}$$

The substitution $u = \alpha^{\frac{1}{3}}$, $d\alpha = 3u^2\, du$ yields:

$$f(1) = \frac{2\pi}{\sqrt{3}} \int_0^1 \frac{u}{u^2 + u + 1}\, du$$

$$= \frac{2\pi}{\sqrt{3}} \left[\int_0^1 \frac{2u + 1}{2(u^2 + u + 1)}\, du - \int_0^1 \frac{du}{2(u^2 + u + 1)} \right]$$

Both integrals can be easily evaluated (The first by the substitution $y = u^2 + u + 1$, $dy = (2u + 1)\, du$ and the second by completing the square). Therefore,

$$f(1) = \frac{2\pi}{\sqrt{3}} \left[\frac{\ln 3}{2} - \frac{\pi\sqrt{3}}{18} \right]$$

$$= \frac{\pi\sqrt{3}\ln 3}{3} - \frac{\pi^2}{9}$$

Example 3: Evaluate $\displaystyle\int_{-\frac{\pi}{2}}^{\frac{\pi}{2}} \frac{\ln(1 + \alpha \sin x)}{\sin x}\, dx$ for $|\alpha| < 1$.

Solution
Define

$$f(\alpha) = \int_{-\frac{\pi}{2}}^{\frac{\pi}{2}} \frac{\ln(1 + \alpha \sin x)}{\sin x}\, dx$$

We then have

3.2. DIRECT APPROACH

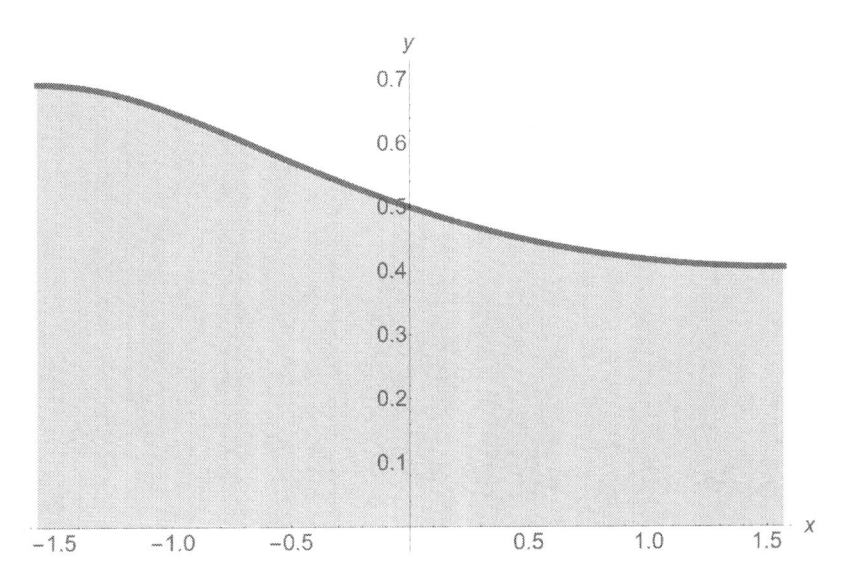

Figure 3.3: Case when $\alpha = .5$

$$f'(\alpha) = \int_{-\frac{\pi}{2}}^{\frac{\pi}{2}} \frac{1}{1+\alpha \sin x} dx$$

A standard substitution here is the **half angle tangent substitution**, $u = \tan\left(\frac{x}{2}\right)$, $dx = \frac{2}{u^2+1} du$. This lets us express $\sin x$ as

$$\sin(x) = \frac{2u}{1+u^2}$$

Thus,

$$f'(\alpha) = \int_{-1}^{1} \frac{2}{u^2 + 2\alpha u + 1} du$$

$$= 2 \int_{-1}^{1} \frac{du}{(\sqrt{1-\alpha^2})^2 + (u+\alpha)^2}$$

This integral is in the form:

$$\int_a^b \frac{dx}{x^2 + d^2} = \left[\frac{1}{d} \arctan\left(\frac{x}{d}\right)\right]_a^b$$

Therefore,

$$f'(\alpha) = 2 \int_{-1}^1 \frac{du}{(\sqrt{1-\alpha^2})^2 + (u+\alpha)^2}$$

$$= \left[\frac{2}{\sqrt{1-\alpha^2}} \arctan\left(\frac{u+\alpha}{\sqrt{1-\alpha^2}}\right)\right]_{-1}^1$$

After a tedious round of algebra and simplification, we obtain:

$$f'(\alpha) = \frac{\pi}{\sqrt{1-\alpha^2}}$$

Thus,

$$f(\alpha) = \int f'(\alpha) \, d\alpha$$

$$= \int \frac{\pi}{\sqrt{1-\alpha^2}} \, d\alpha$$

$$= \pi \arcsin \alpha + C$$

Notice that

$$f(0) = \int_{-\frac{\pi}{2}}^{\frac{\pi}{2}} \frac{\ln 1}{\sin x} dx = 0$$

$$\implies C = 0$$

Therefore,

3.2. DIRECT APPROACH

$$f(\alpha) = \int_{-\frac{\pi}{2}}^{\frac{\pi}{2}} \frac{\ln(1 + \alpha \sin x)}{\sin x} dx = \pi \arcsin \alpha$$

Example 4: Find $\int_{1}^{\infty} \frac{\ln(\ln x)}{x^{\alpha}} dx$ for $\alpha > 1$.

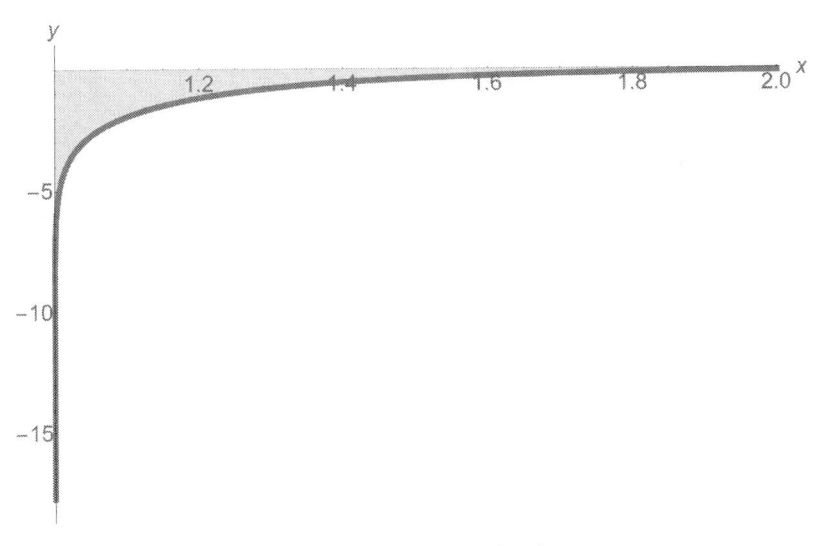

Figure 3.4: Graph of $y = \frac{\ln(\ln x)}{x^{\alpha}}$ for $\alpha = 2$

Solution

Let

$$f(\alpha) = \int_{1}^{\infty} \frac{\ln(\ln x)}{x^{\alpha+1}} dx$$

Substituting $y = \ln x, dy = \frac{dx}{x}$,

$$f(\alpha) = \int_0^\infty \frac{\ln y}{e^{\alpha y}} dy$$

We used $\alpha + 1$ instead of α to make the aforementioned substitution much cleaner. This is one of many examples where the parameter choice is not immediately obvious, and some vision is required. Now, we can differentiate under the integral sign to get

$$f'(\alpha) = -\int_0^\infty y e^{-\alpha y} \ln y \, dy$$

By applying IBP with $u = -y \ln y$, $dv = e^{-\alpha y} dy$ we get:

$$f'(\alpha) = \left[\frac{y \ln y e^{-\alpha y}}{\alpha}\right]_0^\infty - \frac{1}{\alpha}\int_0^\infty e^{-\alpha y} \ln y + e^{-\alpha y} dy$$

We first need to evaluate the first expression:

$$B = \left[\frac{y \ln y e^{-\alpha y}}{\alpha}\right]_0^\infty = \lim_{y \to \infty}\left[\frac{y \ln y}{e^{\alpha y}\alpha}\right] - \lim_{y \to 0}\left[\frac{\ln y}{\frac{e^{\alpha y}\alpha}{y}}\right]$$

This is an $\frac{\infty}{\infty}$ case in both limits, so we can apply L'Hopital's rule:

$$B = \lim_{y \to \infty}\frac{\ln y + 1}{\alpha^2 e^{\alpha y}} - \lim_{y \to 0}\frac{\frac{1}{y}}{\frac{\alpha^2 y e^{\alpha y} - \alpha e^{\alpha y}}{y^2}}$$

$$= \lim_{y \to \infty}\frac{\ln y + 1}{\alpha^2 e^{\alpha y}} - \lim_{y \to 0}\frac{y}{\alpha^2 y e^{\alpha y} - \alpha e^{\alpha y}}$$

3.2. DIRECT APPROACH

The latter limit can be easily evaluated to equal 0, but the first limit needs one more round of L'Hopital's rule as it is still a $\frac{\infty}{\infty}$ situation:

$$B = \lim_{y \to \infty} \frac{\frac{1}{y}}{\alpha^3 e^{\alpha y}}$$

$$= \lim_{y \to \infty} \frac{1}{\alpha^3 y e^{\alpha y}}$$

$$= 0$$

Therefore,

$$f'(\alpha) = 0 - \frac{1}{\alpha} \int_0^\infty e^{-\alpha y} \ln y + e^{-\alpha y} dy$$

$$= -\frac{1}{\alpha} \left(\int_0^\infty e^{-\alpha y} \ln y \, dy + \int_0^\infty e^{-\alpha y} \, dy \right)$$

The first integral is simply $f(\alpha)$ and the latter integral is a standard integral. We then get the differential equation

$$f'(\alpha) = -\frac{f(\alpha)}{\alpha} - \frac{1}{\alpha^2}$$

We can rearrange the above differential equation and multiply both sides by α to obtain:

$$\alpha f'(\alpha) + f(\alpha) = -\frac{1}{\alpha}$$

$$\frac{d}{d\alpha} \alpha f(\alpha) = -\frac{1}{\alpha}$$

Integrating both sides,

$$\alpha f(\alpha) = -\ln \alpha + C$$

$$f(\alpha) = -\frac{\ln \alpha + C}{\alpha}$$

Since
$$f(\alpha) = \int_0^\infty e^{-\alpha y} \ln y \, dy$$
A standard value would be $f(1) = -\gamma$. Thus,
$$f(\alpha) = -\frac{\ln \alpha + \gamma}{\alpha}$$
$$\therefore \int_1^\infty \frac{\ln(\ln x)}{x^\alpha} dx = f(\alpha - 1) = -\frac{\ln(\alpha - 1) + \gamma}{\alpha - 1}$$

Example 5: Evaluate $\int_0^\infty e^{-x^2} \cos(5x) \, dx$

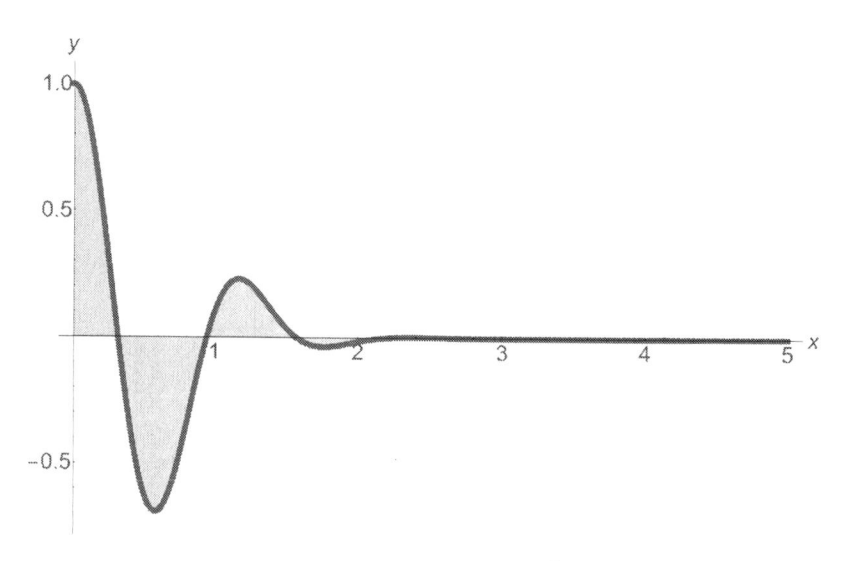

Figure 3.5: Graph of $y = e^{-x^2} \cos(5x)$

Solution

3.2. DIRECT APPROACH

Define a function:

$$f(\alpha) = \int_0^\infty e^{-x^2} \cos(\alpha x) \, dx$$

Differentiating under the integral sign,

$$f'(\alpha) = -\int_0^\infty x e^{-x^2} \sin(\alpha x) \, dx$$

We can now use IBP with $u = \sin(\alpha x)$, $dv = xe^{-x^2} \, dx$ to obtain:

$$f'(\alpha) = \underbrace{\frac{1}{2}\left[e^{-x^2} \sin(\alpha x)\right]_0^\infty}_{=0-0=0} - \frac{\alpha}{2} \underbrace{\int_0^\infty e^{-x^2} \cos(\alpha x) \, dx}_{f(\alpha)}$$

Notice that we can set up a simple differential equation:

$$f'(\alpha) = -\frac{\alpha}{2} f(\alpha)$$

We can now transition to using Leibniz notation to make solving this a little easier to digest.

$$\frac{df}{d\alpha} = -\frac{\alpha}{2} f$$

Multiplying both sides by $\dfrac{d\alpha}{f}$,

$$\frac{df}{f} = -\frac{\alpha}{2} d\alpha$$

$$\int \frac{\mathrm{d}f}{f} = -\int \frac{\alpha}{2} \mathrm{d}\alpha$$

$$\therefore \ln f = -\frac{\alpha^2}{4} + C$$

Exponentiating both sides,

$$f(\alpha) = e^C e^{-\frac{\alpha^2}{4}}$$

The case when $\alpha = 0$ is half of the famous Gaussian integral:

$$\frac{1}{2}\int_{-\infty}^{\infty} e^{-x^2} \,\mathrm{d}x = \int_{0}^{\infty} e^{-x^2} \,\mathrm{d}x = \frac{\sqrt{\pi}}{2}$$

See (9.9) for proof. This integral is core to many disciplines due to its applications in statistics.

$$\implies f(\alpha) = \frac{\sqrt{\pi}}{2} e^{-\frac{\alpha^2}{4}}$$

To get our desired integral, we evaluate $f(5)$ to get:

$$f(5) = \frac{\sqrt{\pi}}{2} e^{-\frac{5^2}{4}}$$

$$\therefore \int_{0}^{\infty} e^{-x^2} \cos(5x) \,\mathrm{d}x = \frac{\sqrt{\pi}}{2} e^{-\frac{25}{4}}$$

Example 6: Find $\int_{0}^{\frac{\pi}{4}} \frac{\arctan(4\sin(2x))}{\sin(2x)} \mathrm{d}x$

Solution

3.2. DIRECT APPROACH

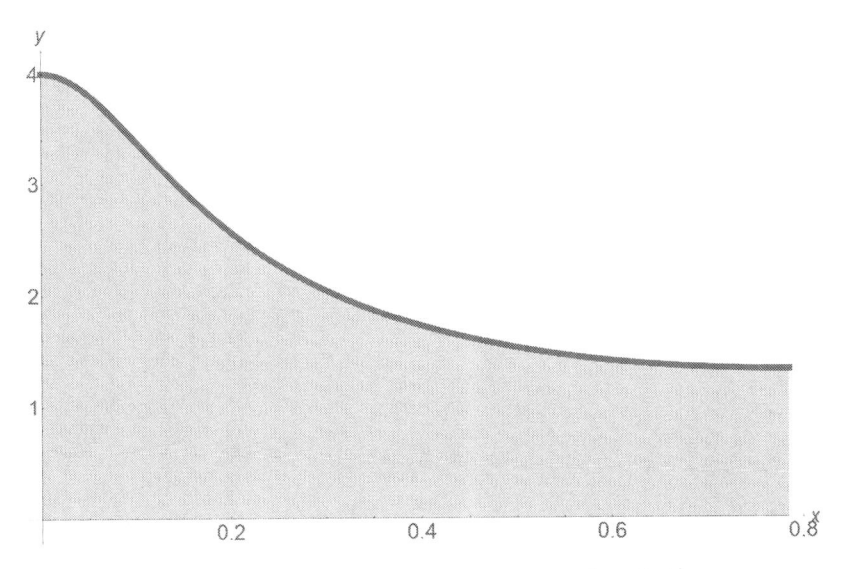

Figure 3.6: Graph of the $y = \frac{\arctan(4\sin(2x))}{\sin(2x)}$

The substitution $u = 2x$, $du = 2\,dx$ gives

$$I = \frac{1}{2}\int_0^{\frac{\pi}{2}} \frac{\arctan(4\sin u)}{\sin u}\,du$$

Now, define a function:

$$f(\alpha) = \int_0^{\frac{\pi}{2}} \frac{\arctan(\alpha \sin u)}{\sin u}\,du$$

Differentiating under the integral sign,

$$f'(\alpha) = \int_0^{\frac{\pi}{2}} \frac{1}{1+\alpha^2 \sin^2 u}\,du$$

We can multiply by $\frac{\sec^2 u}{\sec^2 u}$ to get:

$$f'(\alpha) = \int_0^{\frac{\pi}{2}} \frac{\sec^2 u}{\sec^2 u + \alpha^2 \tan^2 u} du$$

Using the fact that $\sec^2 u = 1 + \tan^2 u$ we can write:

$$f'(\alpha) = \int_0^{\frac{\pi}{2}} \frac{\sec^2 u}{\tan^2 u + 1 + \alpha^2 \tan^2 u} du$$

$$= \int_0^{\frac{\pi}{2}} \frac{\sec^2 u}{(\alpha^2 + 1)\tan^2 u + 1} du$$

$$= \frac{1}{\alpha^2 + 1} \int_0^{\frac{\pi}{2}} \frac{\sec^2 u}{\tan^2 u + \frac{1}{\alpha^2+1}} du$$

The substitution $y = \tan u$, $dy = \sec^2 u \, du$ then gives:

$$f'(\alpha) = \frac{1}{\alpha^2 + 1} \int_0^\infty \frac{dy}{y^2 + \frac{1}{\alpha^2+1}}$$

Which is a standard integral that evaluates to

$$f'(\alpha) = \frac{\pi}{2\sqrt{\alpha^2 + 1}}$$

Integrating both sides,

$$\int f'(\alpha) \, d\alpha = \int \frac{\pi}{2\sqrt{\alpha^2 + 1}} d\alpha$$

$$= \frac{\pi}{2} \sinh^{-1} \alpha + C$$

$$= \frac{\pi}{2} \ln\left(\alpha + \sqrt{\alpha^2 + 1}\right) + C$$

3.2. DIRECT APPROACH

Using $f(0) = 0$ lets us determine the value of C. Hence,

$$f(\alpha) = \frac{\pi}{2} \ln\left(\alpha + \sqrt{\alpha^2 + 1}\right)$$

$$\therefore I = \frac{1}{2}f(4) = \frac{\pi}{4} \ln\left(4 + \sqrt{17}\right)$$

Example 7: Evaluate $\displaystyle\int_0^\infty \int_0^\infty \frac{\sin x \sin y \sin(x+y)}{xy(x+y)} \, dx \, dy$

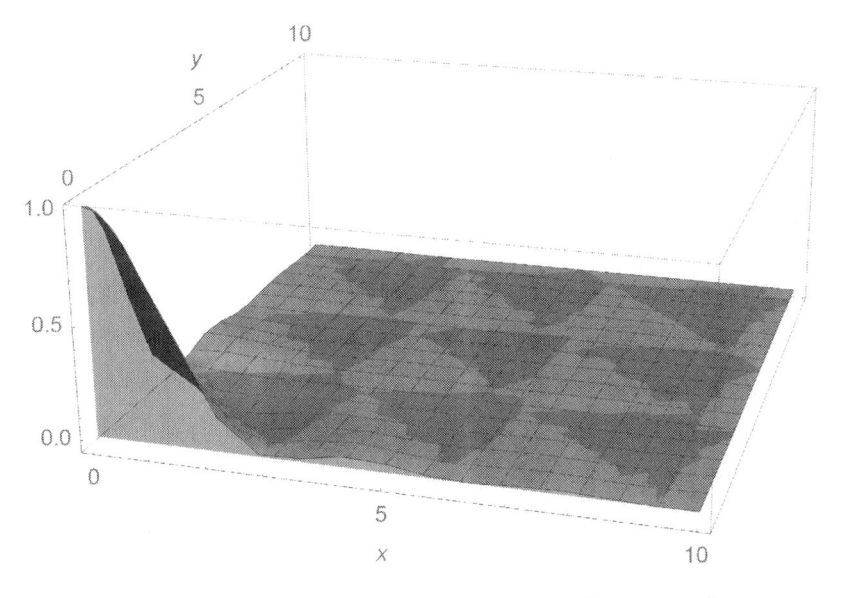

Figure 3.7: Three-dimensional plot of integrand

Solution
Recall that

$$\sin(x + y) = \sin x \cos y + \cos x \sin y$$

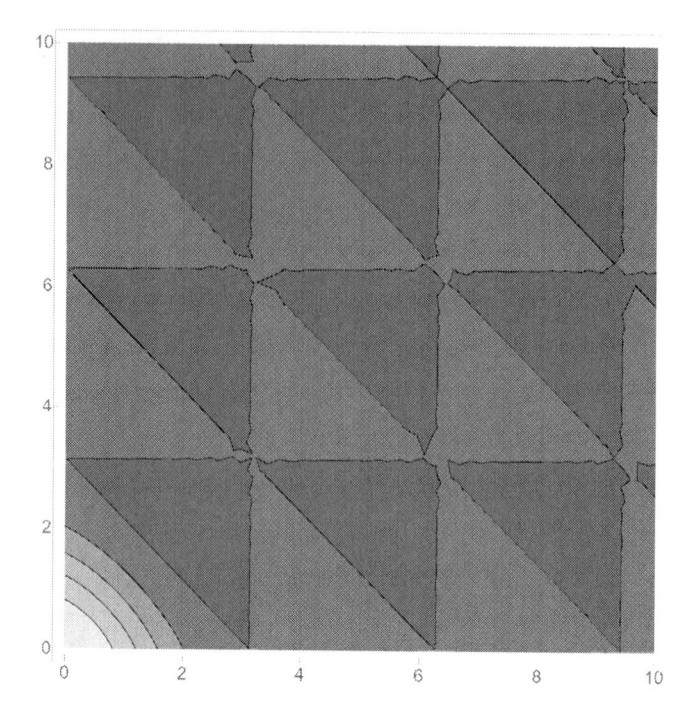

Figure 3.8: Contour plot of integrand

Thus,

$$I = \int_0^\infty \int_0^\infty \frac{(\sin x \cos y + \cos x \sin y)\sin x \sin y}{xy(x+y)} \mathrm{d}x \, \mathrm{d}y$$

By symmetry,

$$I = 2\int_0^\infty \int_0^\infty \frac{\sin^2 x \cos y \sin y}{xy(x+y)} \mathrm{d}x \, \mathrm{d}y$$

Using the fact that $2\sin\theta\cos\theta = \sin 2\theta$, we have:

$$I = \int_0^\infty \int_0^\infty \frac{\sin^2 x \sin 2y}{xy(x+y)} \mathrm{d}x \, \mathrm{d}y$$

3.2. DIRECT APPROACH

We can turn the above double integral into a triple integral:

$$I = \int_0^\infty \int_0^\infty \int_0^\infty \frac{\sin^2 x \sin 2y}{xy} e^{-z(x+y)} \mathrm{d}z \, \mathrm{d}x \, \mathrm{d}y$$

Since

$$\int_0^\infty e^{-ax} \, \mathrm{d}x = \frac{1}{a}$$

We will now introduce an elementary theorem.

Theorem

Let $f(x)$ and $g(y)$ be integrable functions. Then the following equality holds:

$$\int_a^b \int_c^d f(x)g(y)\mathrm{d}y \, \mathrm{d}x = \left(\int_a^b f(x) \, \mathrm{d}x\right)\left(\int_c^d g(y) \, \mathrm{d}y\right)$$

Proof. Note that for any constant k (i.e. independent of x),

$$\int_a^b kf(x) \, \mathrm{d}x = k \int_a^b f(x) \, \mathrm{d}x$$

Also,

$$\int_a^b \int_c^d f(x)g(y)\mathrm{d}y \, \mathrm{d}x = \int_a^b \left(\int_c^d f(x)g(y)\mathrm{d}y\right) \mathrm{d}x$$

$$\int_a^b f(x) \left(\int_c^d g(y)\mathrm{d}y\right) \mathrm{d}x$$

Now, since $\int_c^d g(y)\,\mathrm{d}y$ is a constant, we can write

$$= \left(\int_c^d g(y)\,\mathrm{d}y\right) \cdot \left(\int_a^b f(x)\,\mathrm{d}x\right)$$

□

Using this theorem, we can split the integral above so that we have a product of integrals:

$$I = \int_0^\infty \left(\int_0^\infty \frac{\sin^2 x}{x} e^{-zx}\,\mathrm{d}x\right)\left(\int_0^\infty \frac{\sin 2y}{y} e^{-yz}\,\mathrm{d}y\right)\mathrm{d}z$$

Now, define two functions:

$$f_1(z) = \int_0^\infty \frac{\sin^2 x}{x} e^{-zx}\,\mathrm{d}x$$

$$f_2(z) = \int_0^\infty \frac{\sin 2y}{y} e^{-yz}\,\mathrm{d}y$$

Both these integrals can be easily solved using Feynman's trick (Differentiation under the integral sign). Starting with f_1 we get:

$$f_1'(z) = -\int_0^\infty \sin^2 x\, e^{-zx}\,\mathrm{d}x$$

$$= -\frac{1}{2}\int_0^\infty (1-\cos 2x)\,e^{-zx}\,\mathrm{d}x$$

$$= -\frac{1}{2}\left(\int_0^\infty e^{-zx}\,\mathrm{d}x - \int_0^\infty \cos(2x)\,e^{-zx}\,\mathrm{d}x\right)$$

The second integral can be easily evaluated using IBP. We therefore have:

3.2. DIRECT APPROACH

$$f_1'(z) = \frac{z}{2z^2+8} - \frac{1}{2z}$$

Thus,

$$f_1(z) = \int f_1'(z)\,dz = \frac{1}{2}\int \left(\frac{z}{z^2+4} - \frac{1}{z}\right)dz$$

$$= \frac{1}{2}\left(\frac{1}{2}\ln(z^2+4) - \ln z\right) + C$$

It is easy to see that $f_1(\infty) = 0 \implies C = 0$. We then have

$$f_1(z) = \frac{1}{4}\ln\left(1 + \frac{4}{z^2}\right)$$

Likewise, we can differentiate f_2 to obtain:

$$f_2'(z) = -\int_0^\infty \sin(2y)e^{-yz}\,dy$$

Evaluating the above integral using IBP,

$$f_2'(z) = -\frac{2}{z^2+4}$$

$$\implies f_2(z) = -\int \frac{2}{z^2+4}\,dz = -\arctan\left(\frac{z}{2}\right) + C$$

Using the fact that $f_2(\infty) = 0$, we can determine the value of C:

$$C = \lim_{z \to \infty} \arctan\left(\frac{z}{2}\right)$$

$$C = \frac{\pi}{2}$$

$$\therefore f_2(z) = \frac{\pi}{2} - \arctan\left(\frac{z}{2}\right)$$

Since

$$\arctan\left(\frac{1}{x}\right) + \arctan(x) = \frac{\pi}{2}$$

We have that

$$f_2(z) = \arctan\left(\frac{2}{z}\right)$$

Now, back to our original integral:

$$I = \int_0^\infty \left(\int_0^\infty \frac{\sin^2 x}{x} e^{-zx} \, dx \right) \left(\int_0^\infty \frac{\sin 2y}{y} e^{-yz} \, dy \right) dz$$

$$= \frac{1}{4} \int_0^\infty \ln\left(1 + \frac{4}{z^2}\right) \arctan\left(\frac{2}{z}\right) dz$$

We proceed to substitute $\tan t = \frac{2}{z} \implies dz = -2\csc^2 t \, dt$

$$I = \int_0^{\frac{\pi}{2}} t \csc^2(t) \ln(\sec t) \, dt$$

Using IBP with $u = t \ln(\sec t)$, $dv = \csc^2(t) dt$ gives

$$I = -\left[t \ln(\sec t) \cot t \right]_0^{\frac{\pi}{2}} + \int_0^{\frac{\pi}{2}} \cot t \left[t \tan t + \ln(\sec t) \right] dt$$

$$= \int_0^{\frac{\pi}{2}} t \, dt + \int_0^{\frac{\pi}{2}} \ln(\sec t) \cot t \, dt$$

$$= \frac{\pi^2}{8} + \int_0^{\frac{\pi}{2}} \frac{\ln(\sec t)}{\sec^2 t - 1} \tan t \, dt$$

3.2. DIRECT APPROACH

Substituting $e^u = \sec t$, $dt = \dfrac{du}{\tan t}$ we have:

$$I = \frac{\pi^2}{8} + \int_0^\infty \frac{u}{e^{2u} - 1} \, du$$

By the substitution $2u \to u$, the integral is transformed to:

$$I = \frac{\pi^2}{8} + \frac{1}{4} \int_0^\infty \frac{u}{e^u - 1} \, du$$

By definition (16.2) of the zeta function, we know that:

$$\zeta(s) = \frac{1}{\Gamma(s)} \int_0^\infty \frac{x^{s-1}}{e^x - 1} \, dx$$

Therefore,

$$I = \frac{\pi^2}{8} + \frac{1}{4}\zeta(2)$$

Plugging in $\zeta(2) = \frac{\pi^2}{6}$,

$$I = \frac{\pi^2}{8} + \frac{\pi^2}{24}$$

$$= \frac{\pi^2}{6}$$

Example 8: Evaluate $\displaystyle\int_0^\infty \frac{\arctan(3x)\arctan(2x)}{x^2} \, dx$

Solution

Define a function:

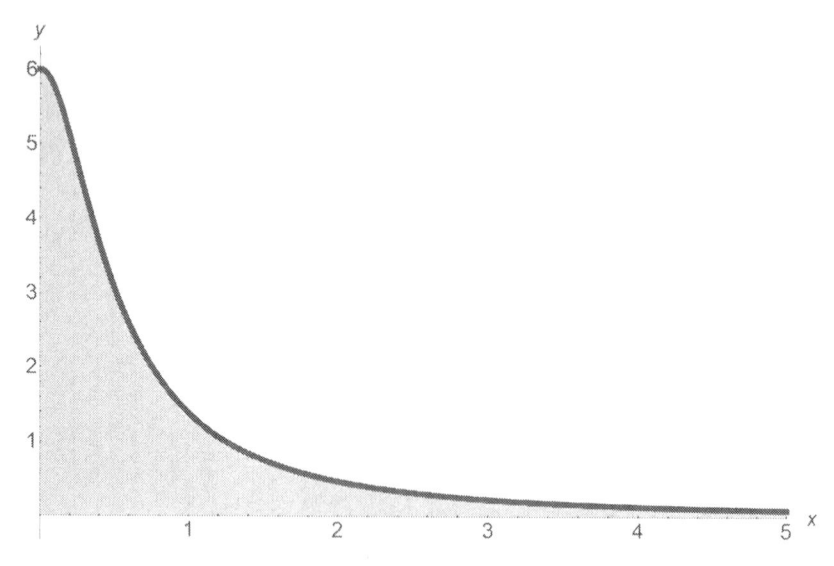

Figure 3.9: Graph of $y = \frac{\arctan(3x)\arctan(2x)}{x^2}$

$$f(\alpha, \beta) = \int_0^\infty \frac{\arctan(\alpha x)\arctan(\beta x)}{x^2} \, dx$$

Since f is a multi-variable function, we will use *partial differentiation*. For the reader who has not taken a course in multi-variable calculus, a partial derivative, $\frac{\partial}{\partial x}$, can be computed by treating all terms not containing x as constants. After differentiating we get

$$\frac{\partial}{\partial \alpha} f(\alpha, \beta) = \int_0^\infty \frac{\arctan(\beta x)}{x(1 + \alpha^2 x^2)} \, dx$$

And,

$$\frac{\partial^2}{\partial \alpha \partial \beta} f(\alpha, \beta) = \int_0^\infty \frac{1}{(1 + \alpha^2 x^2)(1 + \beta^2 x^2)} \, dx$$
$$= \frac{\pi}{2(\alpha + \beta)}$$

3.2. DIRECT APPROACH

The integral above is a standard one, and can be easily calculated using the method of partial fractions. We can now integrate with respect to β to get:

$$\frac{\partial}{\partial \alpha} f(\alpha, \beta) = \int \frac{\partial^2}{\partial \alpha \partial \beta} f(\alpha, \beta) d\beta$$

$$= \int \frac{\pi}{2(\alpha + \beta)} d\beta$$

$$= \frac{\pi}{2} \ln(\alpha + \beta) + C(\alpha)$$

The case $\frac{\partial}{\partial \alpha} f(\alpha, \beta)|_{\beta=0} = 0$ is trivial to calculate. We can then calculate the value of $C(\alpha)$:

$$C = -\frac{\pi}{2} \ln \alpha$$

$$\implies \frac{\partial}{\partial \alpha} f(\alpha, \beta) = \frac{\pi}{2} \left[\ln(\alpha + \beta) - \ln \alpha \right]$$

By symmetry, it is easy to see

$$\frac{\partial}{\partial \beta} f(\alpha, \beta) = \frac{\pi}{2} \left[\ln(\alpha + \beta) - \ln \beta \right]$$

We can now integrate with respect to α to obtain an expression for $f(\alpha, \beta)$.

$$f(\alpha, \beta) = \int \frac{\partial}{\partial \alpha} f(\alpha, \beta) \, d\alpha$$

$$= \frac{\pi}{2} \int \ln(\alpha + \beta) - \ln \alpha \, d\alpha$$

$$= \frac{\pi}{2} \left[\alpha \ln(\alpha + \beta) + \beta \ln(\alpha + \beta) - \alpha \ln \alpha \right] + C(\beta)$$

Where $C(\cdot)$ denotes the integration constant analogue for multivariable functions. Similarly,

$$f(\alpha, \beta) = \int \frac{\partial}{\partial \beta} f(\alpha, \beta) \mathrm{d}\beta$$

$$= \frac{\pi}{2} \int \ln(\alpha + \beta) - \ln \beta \; \mathrm{d}\beta$$

$$= \frac{\pi}{2} \left[\alpha \ln(\alpha + \beta) + \beta \ln(\alpha + \beta) - \beta \ln \beta\right] + C(\alpha)$$

Therefore,

$$f(\alpha, \beta) = \frac{\pi}{2} \left[\alpha \ln(\alpha + \beta) + \beta \ln(\alpha + \beta) - \beta \ln \beta - \alpha \ln \alpha\right]$$

Converting into a single logarithm,

$$f(\alpha, \beta) = \frac{\pi}{2} \ln \left(\frac{(\alpha + \beta)^{\alpha+\beta}}{\alpha^\alpha \beta^\beta}\right)$$

Plugging in our values, we have

$$\int_0^\infty \frac{\arctan(3x) \arctan(2x)}{x^2} \mathrm{d}x = \frac{\pi}{2} \ln \left(\frac{5^5}{3^3 \cdot 2^2}\right)$$

3.3 Indirect Approach

Quite a few integrals are actually derivatives of integrals that are easier to compute. One advantage of differentiation under

3.3. INDIRECT APPROACH

the integral sign is that it can go both ways, i.e. one can integrate a derivative of an expression to obtain a result, or one can take the derivative of a known integral to obtain a result. We can start off with an easy example:

Example 9: $\int_0^\infty \frac{x^\alpha \ln x}{(1+x^\alpha)^2} dx$

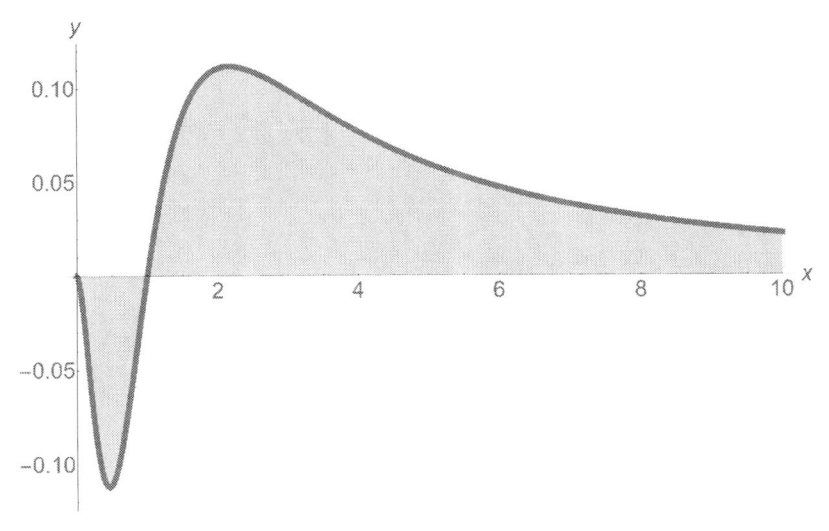

Figure 3.10: Graph of $y = \frac{x^\alpha \ln x}{(1+x^\alpha)^2}$ for $\alpha = 2$

Solution
Define

$$f(\alpha) = \int_0^\infty \frac{dx}{1+x^\alpha}$$

Differentiating under the integral sign gives

$$f'(\alpha) = -\int_0^\infty \frac{x^\alpha \ln x}{(1+x^\alpha)^2} dx$$

Notice that our desired integral is $-f'(\alpha)$. We can now use the result from (2.8),

$$f(\alpha) = \frac{\pi}{\alpha} \csc\left(\frac{\pi}{\alpha}\right) \qquad (3.3)$$

Therefore,

$$I = -f'(\alpha) = \frac{\pi \csc\left(\frac{\pi}{\alpha}\right)}{\alpha^2} - \frac{\pi^2 \csc\left(\frac{\pi}{\alpha}\right) \cot\left(\frac{\pi}{\alpha}\right)}{\alpha^3}$$

Example 10: Evaluate $\displaystyle\int_0^{\frac{\pi}{2}} \sin x \cos x \sqrt{\tan x} \ln(\tan x) \, dx$

Solution

A straightforward substitution would be $u = \sqrt{\tan x}$, $du = \frac{\sec^2 x}{2\sqrt{\tan x}} dx$, but our integral needs some rearranging first. Notice that we can rearrange our desired integral as:

$$I = \int_0^{\frac{\pi}{2}} \frac{4 \sec^2 x \tan^2 x \ln\left(\sqrt{\tan x}\right)}{2 \sec^4 x \sqrt{\tan x}} dx$$

Therefore the substitution $u = \sqrt{\tan x}$ yields:

$$I = 4 \int_0^\infty \frac{u^4 \ln u}{(u^4+1)^2} du$$

3.4. EXERCISE PROBLEMS 131

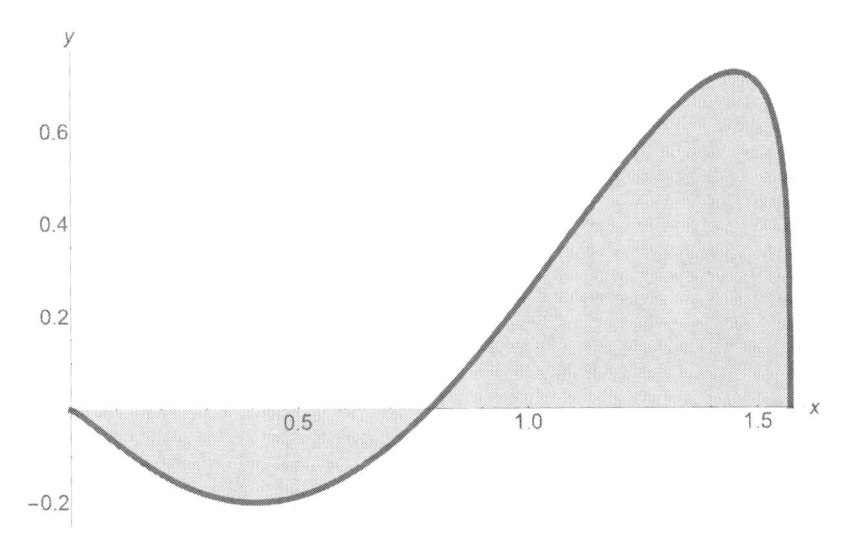

Figure 3.11: Graph of $y = \sin x \cos x \sqrt{\tan x} \ln(\tan x)$

This integral has been already evaluated (See (3.3)). Thus,

$$I = 4\left(\frac{\pi}{4^2}\csc\left(\frac{\pi}{4}\right)\right)\left(1 - \frac{\pi}{4}\cot\left(\frac{\pi}{4}\right)\right)$$

$$= \frac{\pi\sqrt{2}(4-\pi)}{16}$$

3.4 Exercise Problems

1) Find the value of $\displaystyle\int_0^1 \frac{x^2-1}{\ln x}\,\mathrm{d}x$

2) Evaluate $\int_0^1 \dfrac{\ln(1+x)}{1+x^2}dx$

3) Find $\int_0^{\pi/2} \ln\left(\alpha\cos^2 x + \beta\sin^2 x\right)dx$ for $\alpha, \beta > 0$

4) Use differentiation under the integral sign to evaluate

$$\int_{-\pi}^{\pi} \dfrac{\cos^2 x}{1+a^x}dx$$

For $a > 0$.

5) Evaluate $\int_0^{2\pi} e^{\cos x}\cos(\sin x)\,dx$

6) Prove that $\int_{-\infty}^{\infty} e^{-x^2}dx = \sqrt{\pi}$ (Hint: Let $f(\alpha) = \int_0^{\infty} \dfrac{e^{-\alpha^2(1+x^2)}}{1+x^2}dx$)

7) Find $\int_0^{\pi/2} \dfrac{x}{\tan x}dx$

8) Find $\int_0^{\infty} \dfrac{\cos x}{1+x^2}dx$ (Hint: Use the indirect method)

9) Evaluate $\int_0^{\pi/2} \dfrac{dx}{\left(\alpha\cos^2 x + \beta\sin^2 x\right)^2}$ for $\alpha, \beta > 0$.

10) Prove Frullani's theorem

3.4. EXERCISE PROBLEMS

> **Theorem**
>
> Let $a, b > 0$ and let f be continuously differentiable on $[0, \infty)$. Suppose that
> $$\lim_{x \to \infty} f(x)$$
> Exists and is finite. Then the following equality holds:
> $$\int_0^\infty \frac{f(ax) - f(bx)}{x} \, dx = \left(f(\infty) - f(0) \right) \cdot \ln\left(\frac{a}{b}\right)$$

Chapter 4

Sums of Simple Series

4.1 Introduction

In this chapter, we will explore the evaluation of simple series. We will start with finite series and later invoke the concept of a limit to extend our formulas to infinite series. This will be seen in action as we focus on telescoping series later in the chapter! To introduce our chapter, we can begin with the definition of **summation**.

> **Definition**
>
> Summation is the addition of a sequence of numbers, called addends or summands. The result is their sum or total. A series is similar to a sum, but is often used to refer to summations of infinite sequences.

4.2 Arithmetic and Geometric Series

We will start by a famous example provided by the story of the German mathematician Carl Freidrich Gauss in his mathematics class[1]. One day, Gauss' teacher asked his class to add together all the numbers from 1 to 100, a task the teacher thought would occupy the students for quite some time. He was shocked when young Gauss quickly wrote down the correct answer!

The teacher could not understand how his student had calculated the sum so quickly, but the eight-year old Gauss pointed out that the problem was way simpler than the teacher imagined!

Example 1: Evaluate $S = \sum_{n=1}^{100} n$

---

[1] Waltershausen, W. S. von, Gauß, C. F. (1994). Gauss zum Gedächtnis. Vaduz: Sändig Reprint Verlag.

4.2. ARITHMETIC AND GEOMETRIC SERIES

Solution

Many reading this book are familiar with the formula to solve such a problem and perhaps the background of the aforementioned story as well. We begin by expanding S,

$$S = 1 + 2 + 3 + 4 \cdots + 96 + 97 + 98 + 99 + 100$$

Gauss started by recognizing that one can pair the 1st and the 100th term, the 2nd and the 99th term, and so on to make a sum of 101.

$$S = (1 + 100) + (99 + 2) + (98 + 3) + (97 + 4) + \cdots$$

Now, there are $\frac{100}{2}$ such sums equivalent to 101 in S. Therefore, our answer is:

$$S = 101 \left(\frac{100}{2}\right) = 5050$$

The reader might recognize this as the formula

$$\sum_{n=1}^{k} n = \frac{k(k+1)}{2}$$

Example 2: Evaluate $\sum_{n=0}^{10} 2^n$

Solution

We begin by expanding our sum,

$$S = \sum_{n=0}^{10} 2^n = 1 + 2 + 4 + \cdots + 2^{10}$$

CHAPTER 4. SUMS OF SIMPLE SERIES

Consider multiplying S by 2, or the common ratio.

$$2S = 2 + 4 + 8 + \cdots 2^{11}$$

Subtracting $2S$ from S,

$$-S = 1 - 2^{11}$$

$$S = 2^{11} - 1$$

Example 3: Evaluate $\sum_{n=0}^{k} r^n$ for $r \neq 1$ and finite k.

Solution

We can apply the same logic from our last example. Consider

$$S = \sum_{n=0}^{k} r^n = 1 + r + r^2 + \cdots + r^k \qquad (4.1)$$

Multiplying (4.1) by r gives

$$rS = r + r^2 + \cdots + r^{k+1} \qquad (4.2)$$

As we have done previously, we can subtract (4.2) from (4.1) to get

$$S - rS = 1 - r^{k+1}$$

$$(1-r)S = 1 - r^{k+1}$$

4.2. ARITHMETIC AND GEOMETRIC SERIES

Therefore,
$$\sum_{n=0}^{k} r^n = \frac{1 - r^{k+1}}{1 - r} \qquad (4.3)$$

Example 4: Evaluate $\sum_{n=0}^{\infty} r^n$ for $|r| < 1$.

Solution

By definition,
$$\sum_{n=0}^{\infty} r^n = \lim_{k \to \infty} \sum_{n=0}^{k} r^n$$

Plugging in our result from (4.3),
$$\sum_{n=0}^{\infty} r^n = \lim_{k \to \infty} \frac{1 - r^{k+1}}{1 - r}$$

Since $|r| < 1$, as $k \to \infty$, $r^{k+1} = 0$. Thus,
$$\sum_{n=0}^{\infty} r^n = \frac{1}{1 - r}$$

It is now a good time to introduce some standard results. The sum of the first N squares is:

$$\sum_{n=1}^{N} n^2 = \frac{N(N+1)(2N+1)}{6}$$

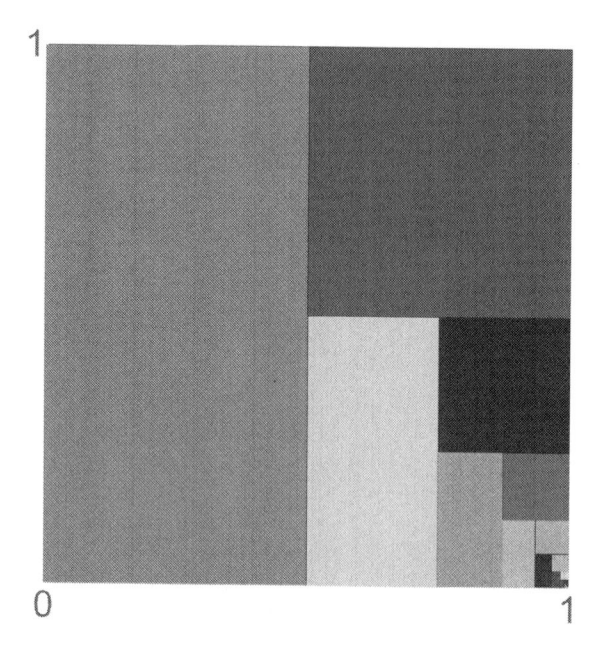

Figure 4.1: A geometric visualization of $\sum_{n=1}^{\infty} \frac{1}{2^n} = 1$. We begin with a rectangle with area $\frac{1}{2}$ and divide it in half with each iteration. The resulting figure is a square of area 1!

Also,

$$\sum_{n=1}^{N} n^3 = \left(\sum_{n=1}^{N} n\right)^2$$

$$= \frac{N^2(N+1)^2}{4}$$

Both formulas can be easily proved by mathematical induction. The proofs for these formulas are left as an exercise for the reader. We also have a general formula for such sums known as Faulhaber's formula[2].

---
[2] Donald E. Knuth (1993). *Johann Faulhaber and sums of powers.* Mathematics of Computation. 61 (203): 277–294. arXiv:math.CA/9207222. doi:10.2307/2152953. JSTOR 2152953.

4.3. ARITHMETIC-GEOMETRIC SERIES

> **Theorem**
>
> In general,
> $$\sum_{n=1}^{N} n^\alpha = \frac{1}{\alpha+1} \sum_{n=0}^{\alpha} \binom{\alpha+1}{n} B_n N^{\alpha+1-n}$$
>
> Where B_n denotes the nth **Bernoulli number**.

> **Definition**
>
> The Bernoulli numbers, usually denoted as B_n, are a sequence of real numbers that satisfy the generating function
> $$\frac{x}{e^x - 1} = \sum_{n=0}^{\infty} \frac{B_n x^n}{n!} \tag{4.4}$$
>
> These numbers have a deep relationship with many of the special functions discussed later in the book. Remarkably, these numbers can also be used to evaluate even zeta function values:
> $$\zeta(2n) = (-1)^{n-1} \frac{2\pi^{2n} B_{2n}}{2(2n)!}$$

4.3 Arithmetic-Geometric Series

We now have formulas for both geometric and arithmetic series, but what about series that are both? For example, the series

$$\sum_{n=1}^{N} n \cdot 4^n$$

This type of series is known is an arithmetic-geometric series. Fortunately, there is a general formula for this series!

Theorem

Let $\{a_n\} = [a + (n-1)b]r^{n-1}$ be defined as an arithmetic-geometric progression. Also, denote

$$S_N = \sum_{n=1}^{N} a_n = a + [a+b]r + [a+2b]r^2 + \cdots + [a+(N-1)b]r^{N-1} \tag{4.5}$$

As the Nth partial sum. We can then express S_N as:

$$S_N = \frac{br(1 - r^{N-1})}{(1-r)^2} + \frac{a - [a + (N-1)b]r^N}{1-r} \tag{4.6}$$

Proof. Multiplying S_N by r gives:

$$rS_N = r\sum_{n=1}^{N} a_n = ar + [a+b]r^2 + [a+2b]r^3 + \cdots + [a+(N-1)b]r^N \tag{4.7}$$

Subtracting (4.7) from (4.5) gives:

$$\begin{array}{rlllll}
S_N = & a+ & [a+b]r + & [a+2b]r^2 + & \cdots + & [a+(N-1)b]r^{N-1} \\
- \quad S_N r = & 0+ & ar + & [a+b]r^2 + & \cdots + & [a+(N-1)b]r^N \\
\hline
(1-r)S_N = a & +br & +br^2 & +\cdots & +br^{N-1} & -[a+(N-1)b]r^N
\end{array}$$

Notice that $br + br^2 + \cdots + br^{N-1}$ is a geometric series. We can therefore write:

$$(1-r)S_N = a + \frac{br(1 - r^{N-1})}{1-r} - [a + (N-1)b]r^N$$

$$\therefore S_N = \frac{br(1 - r^{N-1})}{(1-r)^2} + \frac{a - [a + (N-1)b]r^N}{1-r}$$

4.3. ARITHMETIC-GEOMETRIC SERIES

□

What about infinite arithmetic-geometric series? Before we try to derive the general case, we can get some intuition with an example.

Example 5: Evaluate $\displaystyle\sum_{n=1}^{\infty} \frac{n}{3^n}$

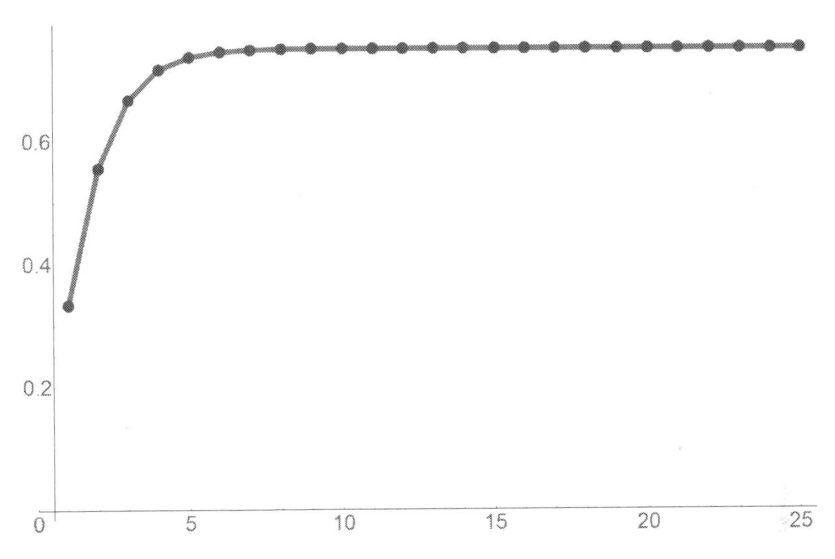

Figure 4.2: Plot of the partial sums

Solution

Consider
$$S = \frac{1}{3} + \frac{2}{9} + \frac{3}{27} + \cdots \tag{4.8}$$

Multiplying S by $\frac{1}{3}$ gives

$$\frac{S}{3} = \frac{1}{9} + \frac{2}{27} + \frac{3}{81} + \cdots \qquad (4.9)$$

Subtracting (4.9) from (4.8),

$$\begin{aligned} S &= \frac{1}{3} + \frac{2}{9} + \frac{3}{27} + \frac{4}{81} + \cdots \\ -\frac{S}{3} &= 0 + \frac{1}{9} + \frac{2}{27} + \frac{3}{81} + \cdots \\ \hline \frac{2}{3}S &= \frac{1}{3} + \frac{1}{9} + \frac{1}{27} + \frac{1}{81} + \cdots \end{aligned}$$

Notice that $\frac{2}{3}S$ is a geometric series with common ratio $\frac{1}{3}$. Hence,

$$\frac{2}{3}S = \sum_{n=1}^{\infty} \frac{1}{3^n} = \frac{\frac{1}{3}}{1 - \frac{1}{3}} = \frac{1}{2}$$

$$\therefore S = \frac{3}{4}$$

Notice that we can also use (4.6) to obtain a general form. We have the following theorem:

Theorem

$$\sum_{n=1}^{\infty} [a + (n-1)b]r^{n-1} = \frac{br}{(1-r)^2} + \frac{a}{1-r} \qquad (4.10)$$

For $|r| < 1$. If $|r| \geq 1$, then:

$$\sum_{n=1}^{\infty} [a + (n-1)b]r^{n-1}$$

Diverges.

4.3. ARITHMETIC-GEOMETRIC SERIES

Proof. Notice that for $|r| \geq 1$, the sequence $\{a_n\} = [a + (n-1)b]r^{n-1}$ does not converge to 0,

$$\lim_{n \to \infty} [a + (n-1)b]r^{n-1} = \infty$$

Therefore, the sum of the terms of this sequence does not converge. However, for $|r| < 1$, the series converges by the **ratio test**, which will be discussed in the next chapter.

To begin, consider the equation for the partial sums of an arithmetic-geometric series,

$$S_N = \sum_{n=1}^{N} [a+(n-1)b]r^{n-1} = \frac{br(1-r^{N-1})}{(1-r)^2} + \frac{a - [a + (N-1)b]r^N}{1-r}$$

By taking the limit as $N \to \infty$, we can obtain an expression for S:

$$S = \lim_{N \to \infty} S_N = \lim_{N \to \infty} \frac{br(1-r^{N-1})}{(1-r)^2} + \frac{a - [a + (N-1)b]r^N}{1-r}$$

$r^N \to 0$ as $N \to \infty$ because we are only considering $|r| < 1$. Also,

$$\lim_{N \to \infty} [a + (N-1)b]r^N = 0$$

Which can be easily shown by L'Hopital's rule. Therefore,

$$S = \frac{br}{(1-r)^2} + \frac{a}{1-r}$$

□

Now, what about any series that is the product of multiple sequences? Is there a general formula? Unfortunately, there is no "general" formula for any series. If that were to exist, many mathematicians would lose their job! However, we can get something almost as good: summation by parts.

4.4 Summation by Parts

By this point of this book, the reader should be familiar with integration by parts, which is proved in Chapter two.

There is a lesser-known discrete analogue to integration by parts called summation by parts, which is for sequences instead of functions.

> **Theorem**
>
> Let $\{a_n\}_{n=1}^{\infty}$ and $\{b_n\}_{n=1}^{\infty}$ be two sequences. Also, denote
>
> $$\mathcal{S}_N = \sum_{n=1}^{N} a_n$$
>
> As the Nth partial sum of the series $\sum_{n=1}^{\infty} a_n$. We then have:
>
> $$\sum_{n=k}^{N} a_n b_n = \mathcal{S}_N b_N - \mathcal{S}_{k-1} b_k - \sum_{n=k}^{N-1} \mathcal{S}_n \left(b_{n+1} - b_n \right) \quad (4.11)$$
>
> This formula is useful not only in the evaluation of series, but also in proving the convergence or divergence of series.

4.4. SUMMATION BY PARTS

Proof. Define
$$S_k^N = \sum_{n=k}^{N} a_n b_n$$

First, notice that:
$$\mathcal{S}_n - \mathcal{S}_{n-1} = a_n$$

Hence,

$$S_k^N = \sum_{n=k}^{N} a_n b_n = \sum_{n=k}^{N} (\mathcal{S}_n - \mathcal{S}_{n-1}) b_n$$

$$= \sum_{n=k}^{N} \mathcal{S}_n b_n - \sum_{n=k}^{N} \mathcal{S}_{n-1} b_n$$

Expanding the expression above

$$\begin{aligned}S_k^N =\ & \mathcal{S}_k b_k + \mathcal{S}_{k+1} b_{k+1} + \mathcal{S}_{k+2} b_{k+2} + \cdots + \underline{\mathcal{S}_N b_N} \\ & - (\underline{\mathcal{S}_{k-1} b_k} + \mathcal{S}_k b_{k+1} + \mathcal{S}_{k+1} b_{k+2} + \cdots + \mathcal{S}_{N-1} b_N)\end{aligned}$$

We can pull the underlined terms to the outside to obtain

$$\begin{aligned}S_k^N =\ & [\mathcal{S}_N b_N - \mathcal{S}_{k-1} b_k] \\ & + \mathcal{S}_k b_k + \mathcal{S}_{k+1} b_{k+1} + \mathcal{S}_{k+2} b_{k+2} + \cdots \mathcal{S}_{N-1} b_{N-1} \\ & - (\mathcal{S}_k b_{k+1} + \mathcal{S}_{k+1} b_{k+2} + \cdots + \mathcal{S}_{N-1} b_N)\end{aligned} \quad (4.12)$$

Notice that combining the expressions from the second and third line in (4.12) gives

$$S = [\mathcal{S}_N b_N - \mathcal{S}_{k-1} b_k] + \mathcal{S}_k(b_k - b_{k+1}) + \mathcal{S}_{k+1}(b_{k+1} - b_{k+2}) + \cdots$$
$$+ \mathcal{S}_{N-1}(b_{N-1} - b_N)$$

$$= [\mathcal{S}_N b_N - \mathcal{S}_{k-1} b_k] + \sum_{n=k}^{N-1} \mathcal{S}_n (b_n - b_{n+1})$$

$$= [\mathcal{S}_N b_N - \mathcal{S}_{k-1} b_k] - \sum_{n=k}^{N-1} \mathcal{S}_n (b_{n+1} - b_n)$$

Hence proved. $\square$

Example 6: Evaluate $\displaystyle\sum_{n=1}^{N} n \cdot 2^n$

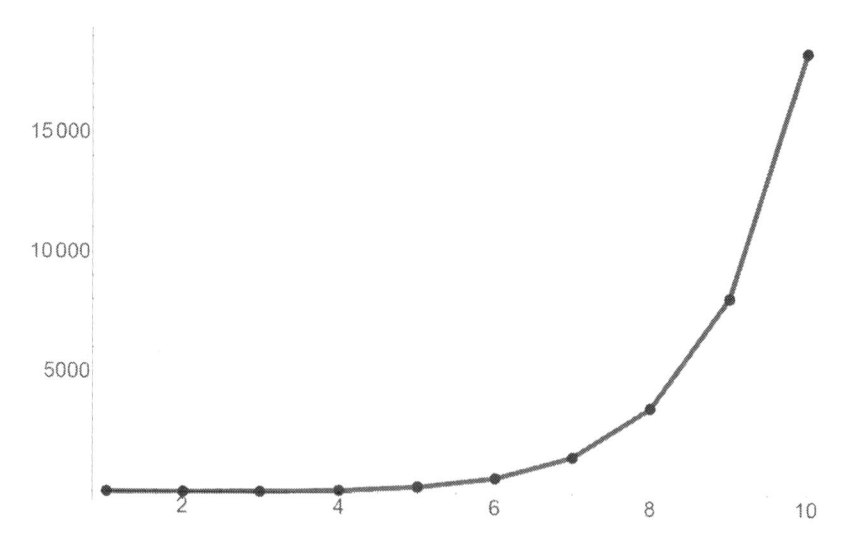

Figure 4.3: Plot of the partial sums when $N = 10$

Solution

4.4. SUMMATION BY PARTS

We will apply (4.11) here,

$$\sum_{n=k}^{N} a_n b_n = \mathcal{S}_N b_N - \mathcal{S}_{k-1} b_k - \sum_{n=k}^{N-1} \mathcal{S}_n (b_{n+1} - b_n)$$

Where

$$\mathcal{S}_N = \sum_{n=1}^{N} a_n$$

Let $a_n = 2^n$ and $b_n = n$. We therefore have:

$$\mathcal{S}_N = \sum_{n=1}^{N} 2^n = 2(2^N - 1)$$

And $\mathcal{S}_0 = 0$. Thus,

$$S_N = \sum_{n=1}^{N} n \cdot 2^n = N \cdot 2(2^N - 1) - \cancel{\mathcal{S}_0 b_1}^{0} - \sum_{n=1}^{N-1} 2(2^n - 1) \cdot (n+1-n)$$

$$= 2N(2^N - 1) - \sum_{n=1}^{N-1} 2(2^n - 1)$$

$$= 2N(2^N - 1) - 2\left[\sum_{n=1}^{N-1} (2^n) - \sum_{n=1}^{N-1} 1\right]$$

$$= 2N(2^N - 1) - 2\left[(2^N - 2) - N + 1\right]$$

$$= N \cdot 2^{N+1} - 2N - 2^{N+1} + 4 + 2N - 2$$

$$= 2^{N+1}(N - 1) + 2$$

Notice that we could have also applied the arithmetic-geometric progression formula in (4.6)!

Example 7: Given a fair standard 6-sided die, what is the expected number of rolls before getting a 6?

Solution

In this problem, we will consider the notion of an "expected value". Although many readers will be familiar with this concept already, it is conveniently defined below:

> **Definition**
>
> Let X be a random variable, or a variable whose value depends on the outcome of a random phenomenon, with a finite number of possible values $x_1, x_2, \cdots, x_N$ with probabilities $p_1, p_2, \cdots, p_N$, respectively. The expected value of X is defined as:
>
> $$\mathbb{E}[X] = \sum_{n=1}^{N} x_n p_n$$
>
> For a random variable X with a countably infinite number of values $x_1, x_2, \cdots$ with probabilities $p_1, p_2, \cdots$, respectively, such that the series $\sum_{n=1}^{\infty} |x_n| p_n$ is convergent, the expected value of X is:
>
> $$\mathbb{E}[X] = \sum_{n=1}^{\infty} x_n p_n \qquad (4.13)$$
>
> The stipulation for the absolute convergence of the series $\sum_{n=1}^{\infty} x_n p_n$ is extremely important because of the Riemann rearrangement theorem (See (5.2)). There is much more to the notion of expected value, but this basic introduction will suffice for our purposes.

Let X be the number of dice rolls till the number 6 is obtained. Notice that the probability that the number 6 is obtained on any given roll of dice is $\frac{1}{6}$ and the probability of getting any number but 6 is $1 - \frac{1}{6} = \frac{5}{6}$.

4.4. SUMMATION BY PARTS

The probability of the first roll yielding the number 6 is $\frac{1}{6}$, but the probability of the second roll yielding the number 6 is $\frac{5}{6} \cdot \frac{1}{6}$, as there is only a $\frac{5}{6}$ chance that the second roll will happen. This is because for the second roll to happen, the first roll must *not* yield a 6. We can continue this pattern such that for the Nth roll,

$$p_N = \frac{1}{6} \cdot \left(\frac{5}{6}\right)^{N-1}$$

Note that X has a countably infinite number of values, which are $1, 2, 3 \cdots$ (1 roll, 2 rolls, etc.). Therefore,

$$\mathbb{E}[X] = 1 \cdot \frac{1}{6} + 2 \cdot \frac{1}{6} \cdot \left(\frac{5}{6}\right)^1 + 3 \cdot \frac{1}{6} \cdot \left(\frac{5}{6}\right)^2 + \cdots$$

$$= \frac{1}{6} \sum_{n=1}^{\infty} n \left(\frac{5}{6}\right)^{n-1}$$

By (4.13). Notice that this an arithmetic-geometric progression, which we have generalized in (4.10):

$$\sum_{n=1}^{\infty} [a + (n-1)b] r^{n-1} = \frac{br}{(1-r)^2} + \frac{a}{1-r}$$

We can evaluate our desired sum by setting $a = 1$, $b = 1$, $r = \frac{5}{6}$. Thus,

$$\mathbb{E}[X] = \frac{1}{6} \sum_{n=1}^{\infty} n \left(\frac{5}{6}\right)^{n-1}$$

$$= \frac{1}{6}(30 + 6)$$

$$= 6$$

This means that the expected number of rolls till one obtains a 6 is six. This is intuitive as rolling the dice a large number of times, say 600 times, will yield about $\frac{600}{6} = 100$ 6's. Since the "gaps" between the occurrences of the number 6 sum up to 600 and there are about 100 6's, the average "gap" is 6 rolls.

4.5 Telescoping Series

> **Definition**
>
> A telescoping series is a series whose partial sums only have a fixed number of terms after cancellation[a]. This method of cancelling terms in partial sums is known as the method of differences. For infinite series, one must take the limit of the partial sums.
>
> More formally, let $\{a_n\}_{n=1}^{\infty}$ be a sequence of real numbers, then
>
> $$\sum_{n=1}^{N} a_n - a_{n-1} = a_N - a_0$$
>
> If $\lim_{n \to \infty} a_n = 0$, we can write:
>
> $$\sum_{n=1}^{\infty} a_n - a_{n-1} = -a_0$$
>
> ---
> [a]Thomson, B. S., Bruckner, J. B., Bruckner, A. M. (2008). *Elementary real analysis*. Upper Saddle River, NJ: Prentice-Hall.

What does this look like? Let's take a dive!

Example 8: Find $\displaystyle\sum_{n=1}^{\infty} \frac{1}{n(n+1)}$

Solution

4.5. TELESCOPING SERIES

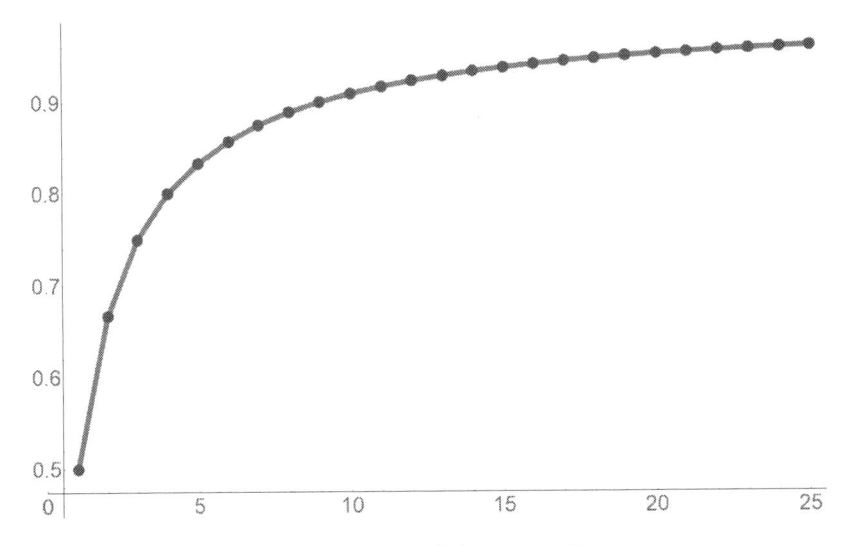

Figure 4.4: Plot of the partial sums

Consider the partial fraction decomposition:

$$\frac{1}{n(n+1)} = \frac{1}{n} - \frac{1}{n+1}$$

We therefore have:

$$S = \sum_{n=1}^{\infty} \frac{1}{n(n+1)} = \sum_{n=1}^{\infty} \left(\frac{1}{n} - \frac{1}{n+1} \right)$$

For telescoping series, it is often helpful to expand the partial sums of the series:

$$S_N = \left(\frac{1}{1} - \frac{\cancel{1}}{\cancel{2}} \right) + \left(\frac{\cancel{1}}{\cancel{2}} - \frac{\cancel{1}}{\cancel{3}} \right) + \left(\frac{\cancel{1}}{\cancel{3}} - \frac{\cancel{1}}{\cancel{4}} \right) + \cdots + \left(\frac{\cancel{1}}{\cancel{N}} - \frac{1}{N+1} \right)$$

$$S_N = 1 - \frac{1}{N+1}$$

Note that S_N denotes the Nth partial sum. Therefore,

$$S = \lim_{N \to \infty} S_N = 1$$

Example 9: Define a sequence $\{F_n\}_{n=0}^{\infty}$ such that the sequence obeys $F_n = F_{n-1} + F_{n-2}$ and $F_0 = F_1 = 1$. This sequence is $1, 1, 2, 3, 5 \cdots$ and is the infamous Fibonacci sequence. Evaluate

$$\sum_{n=1}^{\infty} \frac{1}{F_{n-1}F_{n+1}}$$

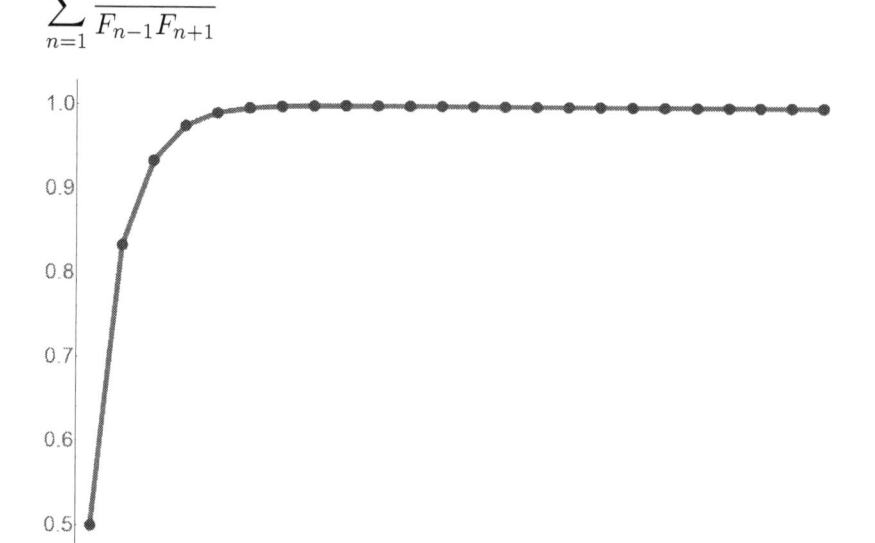

Figure 4.5: Plot of the partial sums. Notice that the series converges very fast

Solution

Consider multiplying the summand by $\dfrac{F_n}{F_n}$,

$$S = \sum_{n=1}^{\infty} \frac{F_n}{F_{n-1}F_n F_{n+1}}$$

Since $F_{n+1} = F_n + F_{n-1} \implies F_n = F_{n+1} - F_{n-1}$, we can write:

4.5. TELESCOPING SERIES

$$S = \sum_{n=1}^{\infty} \frac{F_{n+1} - F_{n-1}}{F_{n-1} F_n F_{n+1}}$$

By partial fraction decomposition,

$$S = \sum_{n=1}^{\infty} \left(\frac{1}{F_n F_{n-1}} - \frac{1}{F_n F_{n+1}} \right)$$

Now, consider the partial sums of S,

$$S_N = \left(\frac{1}{F_0 F_1} - \frac{1}{F_1 F_2} \right) + \left(\frac{1}{F_1 F_2} - \frac{1}{F_2 F_3} \right) + \cdots$$

$$+ \left(\frac{1}{F_{N-1} F_N} - \frac{1}{F_N F_{N+1}} \right)$$

$$= \frac{1}{F_0 F_1} - \frac{1}{F_N F_{N+1}} = 1 - \frac{1}{F_N F_{N+1}}$$

Taking the limit gives

$$S = \lim_{N \to \infty} S_N = 1$$

Example 10: Define

$$S_N = \frac{1}{4^2 - 4} + \frac{1}{6^2 - 4} + \cdots + \frac{1}{(2N)^2 - 4}$$

Find S_{10} and $\lim_{N \to \infty} S_N$.

Solution

As always, we will try to use the method of partial fraction decomposition for problems like these. Since the problem is asking

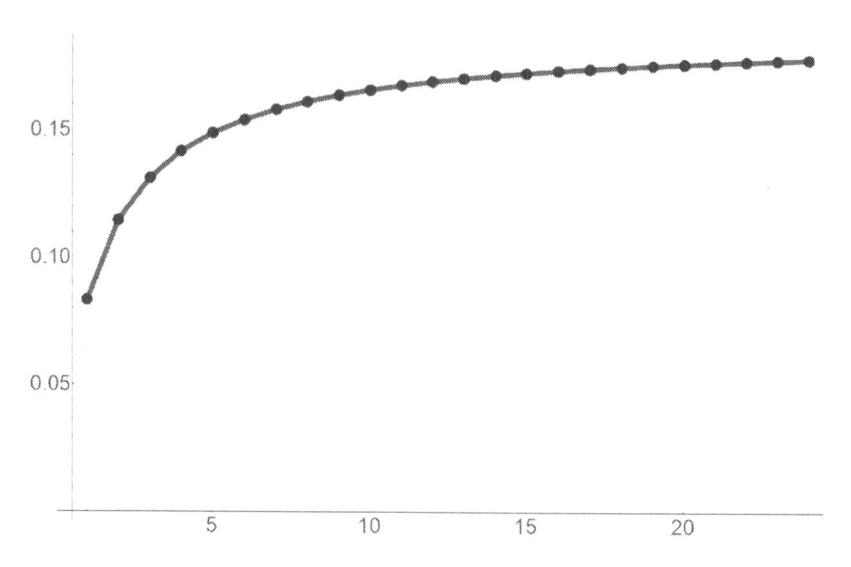

Figure 4.6: Plot of the partial sums

for *both* the evaluation of S_N for some arbitrary N as well as $\lim_{N \to \infty} S_N$, it would be wise to derive a general expression for any S_N. We begin by expressing S_N as:

$$S_N = \sum_{n=2}^{N} \frac{1}{(2n)^2 - 4}$$

$$= \sum_{n=2}^{N} \frac{1}{(2n-2)(2n+2)}$$

$$= \sum_{n=2}^{N} \frac{1}{4}\left(\frac{1}{2n-2} - \frac{1}{2n+2}\right)$$

$$= \frac{1}{8}\sum_{n=2}^{N}\left(\frac{1}{n-1} - \frac{1}{n+1}\right)$$

4.5. TELESCOPING SERIES

$$= \frac{1}{8}\left[\left(\frac{1}{1} - \frac{\cancel{1}}{\cancel{3}}\right) + \left(\frac{1}{2} - \frac{\cancel{1}}{\cancel{4}}\right) + \left(\frac{\cancel{1}}{\cancel{3}} - \frac{\cancel{1}}{\cancel{5}}\right) + \left(\frac{\cancel{1}}{\cancel{4}} - \frac{\cancel{1}}{6}\right) + \cdots\right.$$

$$\left. + \left(\frac{1}{N-1} - \frac{1}{N+1}\right)\right.$$

It is easy to see that this sum telescopes, as only the terms $\frac{1}{1}, \frac{1}{2}, -\frac{1}{N}$, and $-\frac{1}{N+1}$ remain. Therefore,

$$S_N = \frac{1}{8}\left(\frac{3}{2} - \frac{1}{N} - \frac{1}{N+1}\right)$$

$$= \frac{1}{8}\left(\frac{3}{2} - \frac{2N+1}{N(N+1)}\right)$$

$$\implies S_{10} = \frac{1}{8}\left(\frac{3}{2} - \frac{21}{10 \cdot 11}\right) = \frac{9}{55} \approx 0.164$$

And,

$$\lim_{N \to \infty} S_N = \frac{1}{8} \cdot \frac{3}{2} = \frac{3}{16} = .1875$$

Notice that our telescoping sum was not of the form

$$\sum_{n=1}^{\infty} a_n - a_{n-1}$$

Instead, it was of the form

$$\sum_{n=1}^{\infty} a_n - a_{n-2}$$

Since $\frac{1}{n+1}$ and $\frac{1}{n-1}$ are two terms apart. In general, for any constant $k \in \mathbb{N}^+$, the series

$$\sum_{n=1}^{\infty} a_n - a_{n-k}$$

Telescopes.

Example 11: Find $\sum_{n=1}^{N} n \cdot n!$

Solution

Notice that:

$$(n+1)! = n!(n+1) = n \cdot n! + n!$$
$$\implies n \cdot n! = (n+1)! - n!$$

Hence,

$$S = \sum_{n=1}^{N} n \cdot n! = \sum_{n=1}^{N} (n+1)! - n!$$

$$= (2! - 1!) + (3! - 2!) + (4! - 3!) + \cdots + \big((N+1)! - N!\big)$$

This sum telescopes,

$$\therefore S = (N+1)! - 1$$

4.6 Trigonometric Series

Many trigonometric series are also telescoping. The strategy in evaluating these series is to utilize one or more trigonometric identities from the list of trigonometric identities in this book.

Example 12:[3] Evaluate $\sum_{n=1}^{\infty} \arctan\left(\dfrac{2}{n^2}\right)$

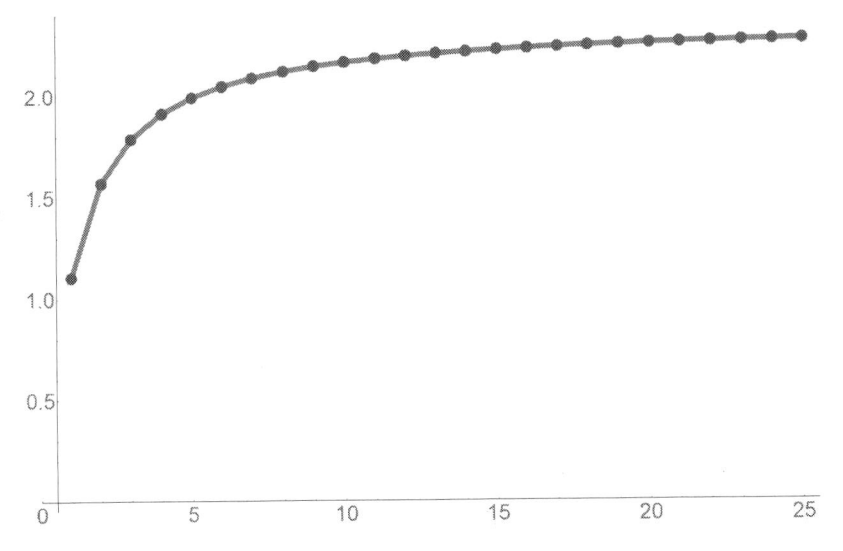

Figure 4.7: Plot of the partial sums

Solution

Consider the trigonometric identity,

$$\tan(x \pm y) = \frac{\tan x \pm \tan y}{1 \mp \tan x \tan y}$$

---

[3] This problem was proposed by Anglesio in 1993: J. Anglesio, Elementary problem 10292, Amer. Math. Monthly 100, (1993), 291. It also appears in J. W. L. Glaisher, A theorem in trigonometry, Quart. J. Math. 15, (1878), and im S. L. Loney, Plane Trigonometry, Part II, Cambridge University Press, Cambridge, 1893.

Substituting $x \to \arctan x$, $y \to \arctan y$ gives:

$$\tan(\arctan x \pm \arctan y) = \frac{x \pm y}{1 \mp xy}$$

Taking the arctangent of both sides,

$$\arctan x \pm \arctan y = \arctan\left(\frac{x \pm y}{1 \mp xy}\right) \mod \pi \quad (4.14)$$

Letting $x = n+1$, $y = n-1$ gives:

$$\arctan(n+1) - \arctan(n-1) = \arctan\left(\frac{2}{n^2}\right)$$

Hence,

$$S = \sum_{n=1}^{\infty} \arctan\left(\frac{2}{n^2}\right)$$

$$= \sum_{n=1}^{\infty} \left(\arctan(n+1) - \arctan(n-1)\right)$$

We have to be careful here, as $\lim_{N \to \infty} \arctan N \neq 0$. By inspecting the partial sums, we have:

$$S_N = \sum_{n=1}^{N} \left(\arctan(n+1) - \arctan(n-1)\right)$$

$$= -\arctan(0) - \arctan(1) + \arctan(N) + \arctan(N+1)$$

And,

$$S = \lim_{N \to \infty} S_N$$

$$= \lim_{N \to \infty} \left[-\arctan(0) - \arctan(1) + \arctan(N) + \arctan(N+1)\right]$$

4.6. TRIGONOMETRIC SERIES

$$= -\frac{\pi}{4} + 2 \lim_{N \to \infty} [\arctan N]$$

$$= \pi - \frac{\pi}{4}$$

$$= \frac{3\pi}{4}$$

Challenge Problem

Prove

$$\sum_{n=1}^{\infty} \arctan\left(\frac{x^2}{n^2}\right) = \frac{\pi}{4} - \arctan\left(\frac{\tanh\left(\frac{\pi x}{\sqrt{2}}\right)}{\tan\left(\frac{\pi x}{\sqrt{2}}\right)}\right)$$

Note that the case $x = \sqrt{2}$ gives:

$$\sum_{n=1}^{\infty} \arctan\left(\frac{2}{n^2}\right) = \frac{\pi}{4} - \arctan\left(\frac{\tanh(\pi)}{\tan(\pi)}\right)$$

$$= \frac{\pi}{4} + \frac{\pi}{2} = \frac{3\pi}{4}$$

Example 13: Evaluate $\displaystyle\sum_{n=0}^{\infty} \arctan\left(\frac{1}{n^2 + n + 1}\right)$

Solution

From (4.14) we have:

$$\arctan\left(\frac{1}{n}\right) - \arctan\left(\frac{1}{n+1}\right) = \arctan\left(\frac{\frac{1}{n} - \frac{1}{n+1}}{1 + \frac{1}{n(n+1)}}\right) \quad \mod \pi$$

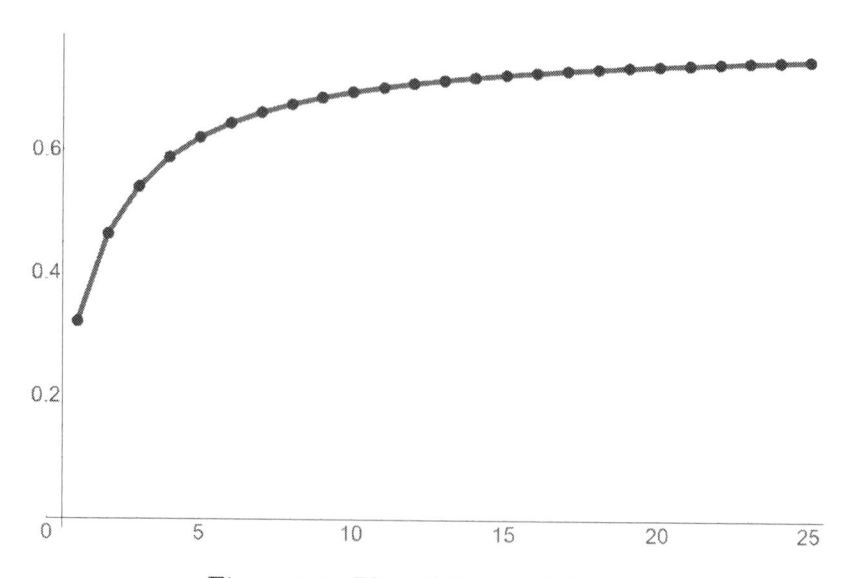

Figure 4.8: Plot of the partial sums

$$= \arctan\left(\frac{1}{n(n+1)\left(1+\frac{1}{n(n+1)}\right)}\right)$$

$$= \arctan\left(\frac{1}{n^2+n+1}\right)$$

Thus,

$$S = \sum_{n=0}^{\infty} \arctan\left(\frac{1}{n^2+n+1}\right)$$

$$= \arctan(1) + \sum_{n=1}^{\infty} \arctan\left(\frac{1}{n^2+n+1}\right)$$

$$= \frac{\pi}{4} + \sum_{n=1}^{\infty}\left(\arctan\left(\frac{1}{n}\right) - \arctan\left(\frac{1}{n+1}\right)\right)$$

$$= \frac{\pi}{4} + \left[\lim_{N\to\infty} \arctan(1) - \arctan\left(\frac{1}{N+1}\right)\right]$$

$$= \frac{\pi}{2}$$

In this chapter, we saw how powerful our algebraic tools are in series. We also got a glimpse into how powerful calculus can be in the evaluation of series through our extensive use of the limit. In the next chapters, we will use the tools we built here to tackle more complex problems.

4.7 Exercise Problems

1) Prove that for any arithmetic series,
$$a_1 + a_2 + \cdots + a_n = \frac{n(a_1 + a_n)}{2}$$

2) Evaluate $\sum_{n=1}^{\infty} \frac{1}{n(n+2)(n+4)}$

3) Find the value of $\sum_{n=1}^{2019} \frac{1 + 2 + 3 + \cdots + n}{1^3 + 2^3 + 3^3 + \cdots + 2019^3}$

4) Find the value of $\sum_{n=1}^{k} n!\,(n^2 + n + 1)$

5) Prove that
$$\sum_{n=1}^{k} \cos n < \frac{1}{2\sin(1/2)} - \frac{1}{2}$$

Using telescoping series (Hint: Multiply the sum by $2\sin(1/2)$ and use the identity $2\sin\alpha\cos\beta = \sin(\alpha+\beta) - \sin(\alpha-\beta)$).

6) Find the value of $\sum_{n=0}^{\infty} \dfrac{\sin^3(3^n)}{3^n}$ (Hint: Use the identity $\sin 3x = 3\sin x - 4\sin^3 x$).

7) Evaluate $\sum_{n=1}^{2019} \ln\left(\dfrac{n}{n^2+3n+2}\right)$ (Hint: Factor and use the properties of logarithms to your advantage).

Part II

Series and Calculus

Chapter 5

Prerequisites

5.1 Introduction

As often as we would like to apply formulas blindlessly, we must be especially careful when dealing with series. Take for example the formula for the geometric series

$$\sum_{n=0}^{\infty} r^n = \frac{1}{1-r} \qquad (5.1)$$

Blindlessly applying the formula above would give the nonsensical result that

$$\sum_{n=0}^{\infty} 2^n = 1 + 2 + 4 + 8 + \cdots = \frac{1}{1-2} = -1$$

This is obviously not true as the series on the LHS is blatantly divergent! The error arises in the fact that (5.1) only holds true for $|r| < 1$, as geometric series only converge for $|r| < 1$. But, how do we define convergence? More specifically, how do we say that a particular series converges to a value S? To invoke rigor into this, we will introduce the following definition:

> **Definition**
>
> A series $S = \sum_{k=1}^{\infty} a_k$ is said to be convergent if the sequence of its partial sums approaches a limit. Equivalently, the series S converges if there exists a number L such that for any arbitrarily small ε there exists some number N such that for all $n \geq N$,
>
> $$|S_n - L| < \varepsilon$$
>
> If L exists, it must be unique, and is the sum of the series.

5.1. INTRODUCTION

To clarify, given an infinite sequence, $\{a_k\}_{k=1}^{\infty}$, the nth partial sum is defined as the sum of the first n terms of the sequence:

$$S_n = \sum_{k=1}^{n} a_k$$

Another territory that is prone to error are conditionally convergent series, or series that do not converge absolutely. We can define such series as follows:

> **Definition**
>
> A series $\sum_{n=0}^{\infty} a_n$ is said to be conditionally convergent if:
>
> $$\lim_{N \to \infty} \sum_{n=0}^{N} a_n = L$$
>
> For some finite L but the series $\sum_{n=0}^{\infty} |a_n|$ diverges.

One example of a conditionally convergent series is

$$\sum_{n=1}^{\infty} \frac{(-1)^n \ln n}{n} = \gamma \ln 2 - \frac{(\ln 2)^2}{2}$$

Another example is given by the well-known alternating harmonic sum:

$$\sum_{n=1}^{\infty} \frac{(-1)^n}{n} = -\ln 2$$

Which converges only conditionally, as the *harmonic series* is divergent.

The harmonic series is given by

$$\sum_{n=1}^{\infty} \frac{1}{n}$$

Many students new to the topic of series find the divergence of this series counterintuitive. This highlights the huge misconception that if $\lim_{n\to\infty} a_n = 0$ then the series $\sum_{n=1}^{\infty} a_n$ converges. This is not true at all! In fact, many well-known divergent series have this property.

Many proofs exist regarding the divergence of the harmonic series, but in this book we will present a simple proof dating back almost 700 years.

Proposition. The harmonic sum is divergent

Proof. This classical proof is due to Nicole Oresme, a well-esteemed medieval philosopher[1]. Oresme considers the sequence $\{H_{2^k}\}_{k=0}^{\infty}$, where $H_k = \sum_{n=1}^{k} \frac{1}{n}$ is the kth harmonic number:

$$H_1 = 1$$

$$H_2 = 1 + \frac{1}{2} = 1 + 1\left(\frac{1}{2}\right)$$

$$H_4 = \left(1 + \frac{1}{2}\right) + \left(\frac{1}{3} + \frac{1}{4}\right) > \left(1 + \frac{1}{2}\right) + \left(\frac{1}{4} + \frac{1}{4}\right)$$

$$= 1 + 2\left(\frac{1}{2}\right)$$

$$\vdots$$

We can easily deduce that $H_{2^k} \geq 1 + k\left(\frac{1}{2}\right)$. Since the subse-

---

[1] Oresme, Nicole (c. 1360). *Quaestiones super Geometriam Euclidis* [Questions concerning Euclid's Geometry].

5.1. INTRODUCTION

quence is unbounded, the harmonic series

$$\sum_{n=1}^{\infty} \frac{1}{n} = \lim_{k \to \infty} H_{2^k}$$

Diverges. The same argument can be extended to any H_{α^k} with $\alpha \in \mathbb{N}$. □

A particularly interesting theorem on conditionally convergent series is the **Riemann series theorem**:

> **Theorem**
>
> The Riemann series theorem, sometimes also called the Riemann rearrangement theorem, states that if an infinite series of real numbers is conditionally convergent, then its terms can be arranged such that the rearranged series converges to an arbitrary real number and can even diverge.

Equivalently, if the series $\sum_{n=1}^{\infty} a_n$ is conditionally convergent, then there exists some *permuation*, given by the function p, such that:

$$\sum_{n=1}^{\infty} a_{p(n)} = L \qquad (5.2)$$

Where $L \in \mathbb{R} \cup \{-\infty, \infty\}$.

Example. Consider the sequence $\{a_n\} = \frac{(-1)^{n-1}}{n}$ and its corresponding series,

$$S = \sum_{n=1}^{\infty} \frac{(-1)^{n-1}}{n}$$

$$= 1 - \frac{1}{2} + \frac{1}{3} - \frac{1}{4} + \cdots = -\ln 2 \qquad (5.3)$$

The value $\ln 2$ comes from the Taylor series of $\ln(1+x)$. Consider rearranging the series as:

$$S_{\text{rearranged}} = \left(1 - \frac{1}{2}\right) - \left(\frac{1}{4}\right) + \left(\frac{1}{3} - \frac{1}{6}\right) - \left(\frac{1}{8}\right) + \cdots$$

Equivalently,

$$S_{\text{rearranged}} = \left(\frac{1}{2k-1} - \frac{1}{2(2k-1)}\right) - \frac{1}{4k}$$

Where $k \in \mathbb{N}$. Notice that all elements of the sequence $\{a_n\}$ are found in $S_{\text{rearranged}}$, but are in a different order. Also,

$$S_{\text{rearranged}} = \frac{1}{2} - \frac{1}{4} + \frac{1}{6} + \cdots$$

$$= \frac{1}{2}\left(1 - \frac{1}{2} + \frac{1}{3} + \cdots\right)$$

$$= \frac{1}{2} \sum_{n=1}^{\infty} \frac{(-1)^{n-1}}{n}$$

$$= \frac{\ln 2}{2}$$

Which is indeed a different value from (5.3)!

5.2 Ways to Prove Convergence

5.2.1 The Comparison Test

Description. Let $\{a_n\}_{n=1}^{\infty}$ and $\{b_n\}_{n=1}^{\infty}$ be two sequences of non-negative numbers. If $a_n \leq b_n$ for all n and $\sum_{n=1}^{\infty} b_n$ converges, then $\sum_{n=1}^{\infty} a_n$ converges. On the other hand, if $a_n \geq b_n$ for all n and $\sum_{n=1}^{\infty} b_n$ diverges, then $\sum_{n=1}^{\infty} a_n$ diverges.

Proof. The proof is trivial and is left as an exercise to the reader. $\square$

5.2.2 The Ratio Test

Description. First published by the French mathematician Jean le Rond d'Alembert, this test is also called d'Alembert's ratio test[2]. Given a sequence of complex numbers $\{a_n\}_{n=1}^{\infty}$ where $a_n \neq 0$ for large n, the ratio test is concerned with the limit:

$$L = \lim_{n \to \infty} \left| \frac{a_{n+1}}{a_n} \right|$$

It states that:

- If $L > 1$, then the series diverges.

- If $L < 1$, then the series converges.

- If $L = 1$, then the ratio test is inconclusive.

---

[2]Zwillinger, D. (Ed.). "Convergence Tests." §1.3.3 in *CRC Standard Mathematical Tables and Formulae, 30th ed.* Boca Raton, FL: CRC Press, p. 32, 1996.

Proof. Although many variations of the ratio test exist, we will present a proof of the original ratio test defined above[3]. Consider:

$$L = \lim_{n \to \infty} \left| \frac{a_{n+1}}{a_n} \right| < 1$$

We will attempt to show convergence by showing that the terms of the sequence $\{a_n\}_{n=1}^{\infty}$ will eventually be less than the terms of a convergent geometric series. Since $L < 1$, there is some number r such that

$$L < r < 1$$

Therefore, for some sufficiently large N, for any $n \geq N$ we have

$$\left| \frac{a_{n+1}}{a_n} \right| < r$$

$$\therefore |a_{n+1}| < r|a_n|$$

Consider repeating this inequality as follows

$$|a_{n+1}| < r\,|a_n|$$
$$|a_{n+2}| < r\,|a_{n+1}| < r^2\,|a_n|$$
$$|a_{n+3}| < r\,|a_{n+2}| < r^3\,|a_n|$$
$$\vdots$$
$$|a_{n+k}| < r\,|a_{n+k-1}| < r^k\,|a_n|$$

Hence for any $k \in \mathbb{N}$

---

[3]Bromwich, T. J. I'A (1908). *An Introduction To The Theory of Infinite Series*. Merchant Books.

5.2. WAYS TO PROVE CONVERGENCE

$$|a_{n+k}| < r^k |a_n| \tag{5.4}$$

Now, consider the series

$$\sum_{k=N+1}^{\infty} a_k = \sum_{k=1}^{\infty} a_{N+k}$$

By (5.4), we can write

$$\sum_{k=1}^{\infty} a_{N+k} < \sum_{k=1}^{\infty} r^k |a_N|$$

Notice that we can write the RHS as

$$\sum_{k=1}^{\infty} r^k |a_N| = |a_N| \sum_{k=1}^{\infty} r^k$$

Which converges since $r < 1$. Therefore, $\sum_{k=1}^{\infty} a_{N+k}$ is convergent by the comparison test. Because

$$\sum_{n=1}^{\infty} a_n = \underline{\sum_{n=1}^{N} a_n} + \sum_{k=1}^{\infty} a_{N+k}$$

And the underlined sum is a finite sum of real numbers, we can conclude that the sum $\sum_{n=1}^{\infty} a_n$ is convergent. Now, to prove the case when $L > 1$ is divergent, recall that:

$$L = \lim_{n \to \infty} \left| \frac{a_{n+1}}{a_n} \right|$$

Because $L > 1$, for sufficiently large N, all $n \geq N$ must satisfy:

$$\left|\frac{a_{n+1}}{a_n}\right| > 1$$

$$|a_{n+1}| > |a_n| \tag{5.5}$$

By (5.5), we know that:

$$\lim_{n \to \infty} |a_n| > 0$$

Because the absolute value of the terms in the sequence $\{a_n\}_{n=1}^{\infty}$ gets larger and larger. Therefore, the series $\sum_{n=1}^{\infty} a_n$ diverges. The special case when $L = 1$ is inconclusive, and can be easily shown through the two sequences:

$$\begin{aligned} \{a_n\}_{n=1}^{\infty} &= \frac{1}{n} \\ \{b_n\}_{n=1}^{\infty} &= \frac{(-1)^n}{n} \end{aligned} \tag{5.6}$$

And their corresponding series,

$$\sum_{n=1}^{\infty} \frac{1}{n} \implies \text{Diverges}$$

$$\sum_{n=1}^{\infty} \frac{(-1)^n}{n} \implies \text{Converges}$$

Since

$$\lim_{n \to \infty} \left|\frac{a_{n+1}}{a_n}\right| = \lim_{n \to \infty} \left|\frac{b_{n+1}}{b_n}\right| = 1$$

□

5.2.3 The Integral Test

Description. Let $f(x)$ be a continuous, positive, monotonically decreasing function on the interval $[b, \infty]$ where $b \in \mathbb{Z}$. Then the infinite series $\sum_{n=b}^{\infty} f(n)$ converges if and only if the integral $\int_{b}^{\infty} f(x) \, dx$ is finite. If the integral diverges, then the corresponding series also diverges.

Proof. Without loss of generality, let $b = 1$. Since we can just shift the index of a series, proving the case when $b = 1$ is sufficient to prove the general case for any $b \in \mathbb{Z}$.

In this proof, we will prove the integral test for *strictly decreasing* $f(x)$. However, the theorem holds true for any monotonically decreasing $f(x)$ (a monotonically decreasing function does not have to be exclusively decreasing, only non-increasing).

Consider the plot below:

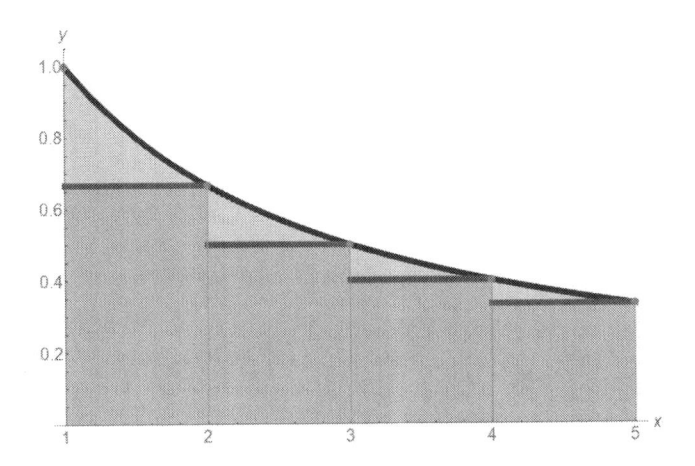

Figure 5.1: Plot of the *right* Riemann sums of an arbitrary strictly decreasing function defined on the interval $[1, \infty]$. The width of the rectangles is a constant 1.

Define a sequence such that $\{a_n\}_{n=1}^{\infty} = f(n)$. Hence,

$$f(2) = a_2, f(3) = a_3, f(4) = a_4 \cdots$$

The approximate area under the curve, or the improper integral on $[1, \infty)$ is:

$$\begin{aligned}
I = \int_1^{\infty} f(x)\, dx &\approx f(2) + f(3) + \cdots \\
&= a_2 + a_3 + a_4 + \cdots \\
&= \sum_{n=2}^{\infty} a_n
\end{aligned}$$

However, this underestimates I as $f(x)$ is *strictly decreasing* on $[1, \infty)$, which implies:

$$\int_N^{N+1} f(x)\, dx > f(N+1) \tag{5.7}$$

And,

$$\int_1^{\infty} f(x)\, dx > \sum_{n=2}^{\infty} f(n) = \sum_{n=2}^{\infty} a_n$$

Since $f(x) > 0$ on $[1, \infty)$, we know that:

$$\int_1^N f(x)\, dx < \int_1^{\infty} f(x)\, dx$$

Thus,

$$\sum_{n=2}^{N} a_n < \int_1^N f(x)\, dx < \int_1^{\infty} f(x)\, dx \tag{5.8}$$

5.2. WAYS TO PROVE CONVERGENCE

By (5.7). To get our desired series, $\sum_{n=1}^{\infty} a_n$, we can add a_1:

$$\sum_{n=1}^{N} a_n < a_1 + \int_{1}^{N} f(x) \, dx = L \qquad (5.9)$$

Now, define the partial sums as

$$\{S_N\} = \sum_{n=1}^{N} a_n$$

By (5.9), we know the sequence $\{S_N\}$ is bounded above by L. Also, since $a_n > 0$, we have that $S_N < S_{N+1}$ for all N. We now know by the monotone convergence theorem that our sequence $\{S_N\}$ converges, and that our corresponding series $\sum_{n=1}^{\infty} a_n$ converges.

> **Theorem**
>
> Informally, the monotone convergence theorem states that if a sequence is increasing and bounded above by a supremum, then the sequence will converge to the supremum. On the other hand, if a sequence is decreasing and is bounded below by an infimum, it will converge to the infimum.

Now, what if our integral diverges? Consider the plot below:

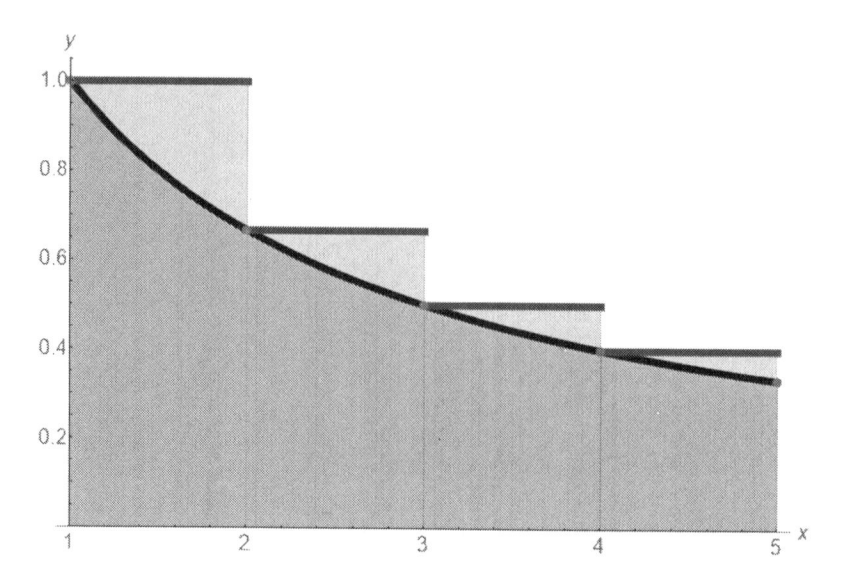

Figure 5.2: Plot of the *left* Riemann sums of an arbitrary monotonically decreasing function defined on the interval $[1, \infty]$. The width of the rectangles is a constant 1.

In this case, the approximate area under the curve on the interval $[1, N]$ given by the Riemann sum is an overestimate. Therefore,

$$\int_1^N f(x) \, \mathrm{d}x < \sum_{n=1}^{N-1} f(n)$$

Since $a_n = f(n)$,

$$\int_1^N f(x) \, \mathrm{d}x < S_{N-1} \qquad (5.10)$$

If the integral diverges, as $N \to \infty$, then

5.2. WAYS TO PROVE CONVERGENCE

$$\int_1^N f(x)\,dx \to \infty$$

(5.10) then shows that $S_N \to \infty$ as well. Since the sequence of partial sums is divergent, so is the corresponding series, $\sum_{n=1}^{\infty} a_n$.

$\square$

5.2.4 The Root Test

Description. Let $\{a_n\}_{n=1}^{\infty}$ be a sequence of real numbers. Define:

$$L = \lim_{n \to \infty} \sqrt[n]{|a_n|}$$

And,

$$S = \sum_{n=1}^{\infty} a_n$$

We then have:

1. If $L < 1$, then the series S is absolutely convergent.

2. If $L > 1$, then the series S diverges.

3. If $L = 1$, then the root test is inconclusive. In other words, S can be absolutely convergent, conditionally convergent, or divergent.

Proof. Consider when $L < 1$. Let r be some number such that $L < r < 1$. Because

$$L = \lim_{n \to \infty} \sqrt[n]{|a_n|}$$

We can write:

$$\sqrt[n]{|a_n|} < r$$

$$\implies r^n > |a_n|$$

For some $n \geq N$ and sufficiently large N. Since $r < 1$, the series:

$$\sum_{n=N}^{\infty} r^n$$

Converges. Because $r^n > |a_n|$ for all $n \geq N$, the series

$$\sum_{n=N}^{\infty} |a_n|$$

Also converges by the comparison test. Therefore, the series

$$\sum_{n=1}^{\infty} a_n = \sum_{n=1}^{N-1} |a_n| + \sum_{n=N}^{\infty} |a_n|$$

Converges since $\sum_{n=1}^{N-1} |a_n|$ is a finite sum of finite terms.

Now, consider the case when $L > 1$. For some sufficiently large N, any $n \geq N$ satisfies:

$$\sqrt[n]{|a_n|} > 1$$

5.2. WAYS TO PROVE CONVERGENCE

$$\implies |a_n| > 1$$

Hence,

$$\lim_{n \to \infty} |a_n| > 0$$

$$\therefore \lim_{n \to \infty} a_n \neq 0$$

We can therefore conclude that the series $\sum_{n=1}^{\infty} a_n$ diverges since the sequence $\{a_n\}_{n=1}^{\infty}$ does not converge to 0.

The case when $L = 1$ is inconclusive, similar to what we saw in the ratio test. To show this, we simply need the two sequences we have provided in (5.6):

$$\{a_n\}_{n=1}^{\infty} = \frac{1}{n}$$

$$\{b_n\}_{n=1}^{\infty} = \frac{(-1)^n}{n}$$

And their corresponding series,

$$\sum_{n=1}^{\infty} \frac{1}{n} \implies \text{Diverges}$$

$$\sum_{n=1}^{\infty} \frac{(-1)^n}{n} \implies \text{Converges}$$

Since

$$\lim_{n\to\infty} \sqrt[n]{|a_n|} = \lim_{n\to\infty} \sqrt[n]{|b_n|} = 1$$

We know that the case $L = 1$ is inconclusive.

□

Note that the root test is stronger than the ratio test.

5.2.5 Dirichlet's Test

The reader equipped with a standard knowledge of calculus should already be familiar with the traditional methods of proving convergence: ratio test, integral test, direct comparison, etc. These methods, although powerful, are sometimes obsolete to series evaluated here. One powerful technique is **Dirichlet's test**. Dirichlet's test, named after the German mathematician Peter Dirichlet, is a powerful method for testing the convergence of series[4].

---

[4]*Demonstration d'un theoreme d'Abel.* Journal de mathematiques pures et appliquees 2nd series, tome 7 (1862), p. 253-255 Archived 2011-07-21 at the Wayback Machine.

> **Theorem**
>
> Dirichlet's test states that if $\{a_n\}_{n=1}^{\infty}$ is a sequence of real numbers and $\{b_n\}_{n=1}^{\infty}$ is a sequence of complex numbers satisfying that:
>
> - $\lim_{n \to \infty} a_n = 0$
> - $a_{n+1} \leq a_n$
> - $\left|\sum_{n=1}^{k} b_n\right| \leq L$ for all $k \in \mathbb{N}$
>
> Where L is a constant, then the series
>
> $$\sum_{n=1}^{\infty} a_n b_n$$
>
> Converges. Notice that if $b_n = (-1)^n$, then Dirichlet's test is simply the alternating series test often encountered in introductory calculus courses.

5.3 Interchanging Summation and Integration

Regarding series and integrals, one must be very careful in interchanging the two. In general, for an infinite sequence of functions $\{f_n\}$,

$$\int \sum_n f_n \, dx \neq \sum_n \int f_n \, dx \tag{5.11}$$

Notice the word "infinite." You might be wondering why finite series get a pass, which is due to the additive property of integrals. However, since ∞ is not a number, when we write an infinite sum we are effectively taking the *limit* as the bound of the sum approaches infinity. Without loss of generality, let $\{f_n\}$ begin at $n = 1$. (5.11) is then effectively:

$$\int \lim_{N \to \infty} \sum_{n=1}^{N} f_n \, dx \neq \lim_{N \to \infty} \sum_{n=1}^{N} \int f_n \, dx$$

It is the interchange of the limit and the integral we are concerned with. Even though in most cases the interchange is valid, one must still be careful. For example,

$$\sum_{n=0}^{\infty} \int_{0}^{2\pi} \sin(x+n) \, dx \qquad (5.12)$$

$$= \sum_{n=0}^{\infty} \left[-\cos(x+n) \right]_{0}^{2\pi}$$

$$\sum_{n=0}^{\infty} \left(\cos(x) - \cos(x+2\pi) \right)$$

$$= \sum_{n=0}^{\infty} 0 = 0$$

Now, consider interchanging summation and integration in (5.12) to give:

$$\int_{0}^{2\pi} \sum_{n=0}^{\infty} \sin(x+n) \, dx$$

Since

$$\sum_{n=0}^{\infty} \sin(x+n)$$

5.3. INTERCHANGING SUMMATION AND INTEGRATION

Diverges for all x, the integral diverges as well. Well, this is contradictory! This example serves as one of many in which the interchange is not valid.

But, what criteria did our example in (5.12) not meet? Instead of investigating each case of (5.11), we can present a general theorem for any sequence of functions $\{f_n\}$ on any interval I. One such theorem is the **Lebesgue dominated convergence theorem**.

> **Theorem**
>
> Let $\{f_n\}_{n=1}^{\infty}$ be a sequence of Lebesgue integrable functions that converge to a limit function f almost everywhere on an interval I. Suppose there exists some Lebesgue integrable function g such that $|f_n| < g$ almost everywhere on I and for all $n \in \mathbb{N}$. Then, f is Lebesgue integrable on I and
>
> $$\lim_{n \to \infty} \int_I f_n(x) \, dx = \int_I \lim_{n \to \infty} f_n(x) \, dx = \int_I f(x) \, dx$$

This theorem is perhaps the most hefty in the book, so let us break it down. First off, what is meant by a "Lebesgue integrable function"?

Although the general definition requires a sizeable amount of measure theory, we will present a definition that will do for all the examples we deal with in this book. We first need to establish the notion of a *measurable* function. In our case, we are trying to employ the Lebesgue measure. A measure can be intuitively understood as the "size" of a set. It is a way of systematically assigning numbers to sets to represent their sizes.

The Lebesgue measure is the measure that coincides most with general notions of "size." For example, the set of all points in $[0, 1]$ has Lebesgue measure equal to its length on the number line, which is 1. Since we are only dealing with intervals, we can simply say that an interval $[a, b]$ has a Lebesgue measure $b - a$. For two dimensional intervals, $[a, b] \times [c, d]$, it coincides with area.

If we are using the Lebesgue definition of measure, then we can say a function f is measurable on I if f is continuous on I.

> **Theorem**
>
> If $f(x)$ is a bounded function defined on some closed interval I such that
> $$\int_I f(x) \, dx$$
> Exists in the Riemann sense, then f is also Lebesgue integrable.

Now, on to the definition of *almost everywhere*. Intuitively, a property that holds almost everywhere is simply just that. Formally, this notion can be expressed by using the concept of measure. If a property holds almost everywhere on an interval I, the *measure* of the set where the property does not hold is 0. This serves us by letting us deal with functions diverging at endpoints.

Now, we will present a series reformulation of the dominated convergence theorem.

5.3. INTERCHANGING SUMMATION AND INTEGRATION

> **Theorem**
>
> Let $\{f_n\}_{n=1}^{\infty}$ be a sequence of Lebesgue integrable functions such that each f_n is nonnegative on I and the sum
> $$\sum_{n=1}^{\infty} f_n(x)$$
> Converges almost everywhere on I to a limit function f. Suppose there exists some Lebesgue integrable function g such that $|f(x)| \leq g(x)$ almost everywhere on I. Then, f is Lebesgue integrable on I and
> $$\int_I f(x)\, dx = \int_I \sum_{n=1}^{\infty} f_n(x)\, dx = \sum_{n=1}^{\infty} \int_I f_n(x)\, dx$$

To see this in action, consider the integral

$$I = \int_0^1 \frac{dx}{1+x}$$

This integral trivially converges to $\ln 2$. But, perhaps we want to substitute the power series expansion

$$\frac{1}{1+x} = \sum_{k=0}^{\infty} (-1)^k x^k$$

To get

$$I = \int_0^1 \sum_{k=0}^{\infty} (-1)^k x^k\, dx$$

$$= \int_0^1 \lim_{n \to \infty} \sum_{k=0}^{n} (-1)^k x^k\, dx$$

Now, is the interchange here justified? Well, all terms of the sequence

$$\{f_n\} = \sum_{k=0}^{n}(-1)^k x^k$$

Are nonnegative. Moreover, the sequence *does* converge almost everywhere on $[0,1]$ to $\dfrac{1}{1+x}$. It does not converge at $x = 1$. Moreover, $f(x)$ is "dominated" by $g(x) = 1$, i.e.

$$\sum_{k=0}^{n}(-1)^k x^k \leq 1$$

Therefore, we are jusified in interchanging the limit (infinite summation) and integral! In doing so we obtain:

$$I = \lim_{n\to\infty} \sum_{k=0}^{n} \int_0^1 (-1)^k x^k \, dx$$

$$= \lim_{n\to\infty} \sum_{k=0}^{n} \frac{(-1)^k}{k+1}$$

Which is the alternating harmonic series that equals the same value of $\ln 2$.

Chapter 6

Evaluating Series

6.1 Introduction

The most widely used series are the power series taught in introductory calculus courses. A classic example is:

$$\frac{1}{1-x} = 1 + x + x^2 + \cdots = \sum_{n=0}^{\infty} x^n \qquad (6.1)$$

Notice that this equation is equivalent to the infinite geometric sum formula. We can do a lot with this series. Try substituting $x \to -x$ to get:

$$\frac{1}{1+x} = 1 - x + x^2 - x^3 + \cdots = \sum_{n=0}^{\infty} (-1)^n x^n \qquad (6.2)$$

We can differentiate both sides to obtain:

$$-\frac{1}{(1+x)^2} = -1 + 2x - 3x^2 + \cdots = \sum_{n=0}^{\infty} n \cdot (-1)^n x^{n-1}$$

We can also integrate! Integrating both sides of (6.1) gives

$$\ln(1-x) = -\sum_{n=0}^{\infty} \frac{x^{n+1}}{n+1}$$

Re-indexing,

$$\ln(1-x) = -\sum_{n=1}^{\infty} \frac{x^n}{n} \qquad (6.3)$$

We can also substitute $x \to -x$ in (6.3) to obtain:

6.2. SOME PROBLEMS

$$\ln(1+x) = \sum_{n=1}^{\infty} \frac{(-1)^{n+1}x^n}{n} \qquad (6.4)$$

It is vital to realize that these powerful tools have their limitations too. It is important to check whether these sums converge before proceeding and using them on a specified interval. For example (6.1), only converges for $|x|<1$. Using (6.1) for $x=5$, for example, would yield nonsensical results.

6.2 Some Problems

Example 1: Evaluate $\sum_{n=1}^{\infty} \frac{n^2}{2^n}$

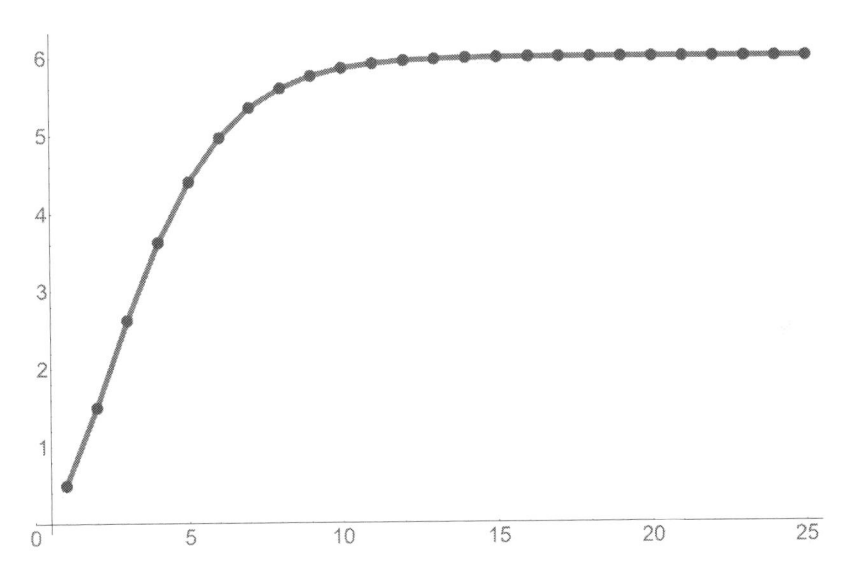

Figure 6.1: Plot of the partial sums

Solution

The series converges by the ratio test. Using (6.1), we have:

$$\frac{x}{1-x} = \sum_{n=1}^{\infty} x^n$$

Differentiating both sides,

$$\frac{1}{(1-x)^2} = \sum_{n=1}^{\infty} nx^{n-1}$$

We can multiply both sides by x to get:

$$\frac{x}{(1-x)^2} = \sum_{n=1}^{\infty} nx^n$$

We can differentiate and multiply by x again to get:

$$\frac{x+x^2}{(1-x)^3} = \sum_{n=1}^{\infty} n^2 x^n$$

Notice that if $x = \frac{1}{2}$, we get our desired sum! Therefore,

$$\sum_{n=1}^{\infty} \frac{n^2}{2^n} = \frac{\frac{1}{2} + \left(\frac{1}{2}\right)^2}{\left(\frac{1}{2}\right)^3}$$

$$= 6$$

Example 2: Find the value of $\sum_{n=1}^{\infty} \frac{1}{2^n \cdot n^2}$

Solution

6.2. SOME PROBLEMS

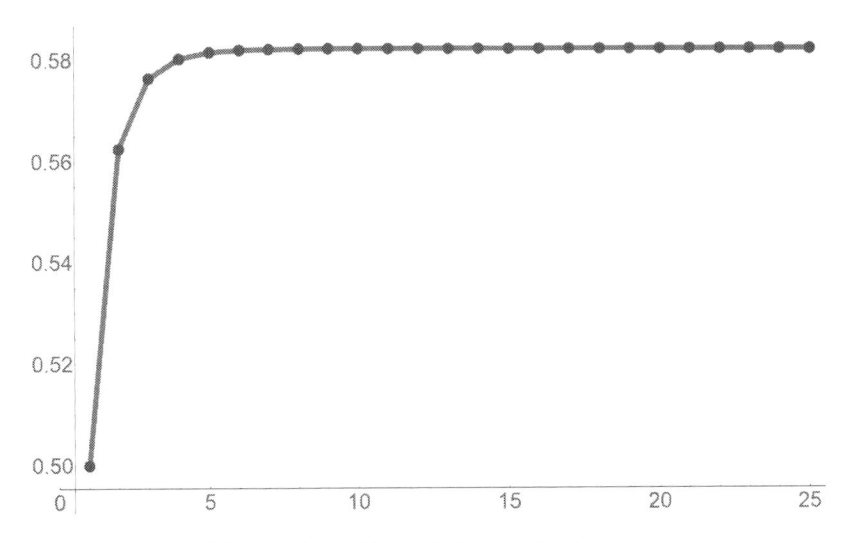

Figure 6.2: Plot of the partial sums

Notice that we can express our desired sum as

$$S = \sum_{n=1}^{\infty} \frac{1}{2^n \cdot n^2}$$

$$= \sum_{n=1}^{\infty} \int_0^{\frac{1}{2}} \frac{x^{n-1}}{n} \, dx$$

Using the dominated convergence theorem, we can switch the order of integration and summation to obtain

$$= \int_0^{\frac{1}{2}} \sum_{n=1}^{\infty} \frac{x^{n-1}}{n} \, dx$$

$$= -\int_0^{\frac{1}{2}} \frac{\ln(1-x)}{x} \, dx$$

Where in the last step we used the power series expansion for $\ln(1-x)$. We can integrate by parts with $u = \ln(1-x)$, $dv = \frac{dx}{x}$ to get

$$S = -\left[\ln x \ln(1-x)\right]_0^{\frac{1}{2}} + \int_0^{\frac{1}{2}} \frac{\ln x}{x-1} dx$$

Since

$$\ln(1+x) = \sum_{n=0}^{\infty} \frac{(-1)^n x^{n+1}}{n+1}$$

We can substitute $x \to x-1$ to get:

$$\ln x = \sum_{n=0}^{\infty} \frac{(-1)^n (x-1)^{n+1}}{n+1}$$

Plugging that series into S,

$$S = -\ln^2\left(\frac{1}{2}\right) + \int_0^{\frac{1}{2}} \sum_{n=0}^{\infty} \frac{(-1)^n (x-1)^n}{n+1} dx$$

We can now use the dominated convergence theorem again,

$$S = -\ln^2\left(\frac{1}{2}\right) + \sum_{n=0}^{\infty} \int_0^{\frac{1}{2}} \frac{(-1)^n (x-1)^n}{n+1} dx$$

$$= -\ln^2\left(\frac{1}{2}\right) + \sum_{n=0}^{\infty} \frac{(-1)^n}{n+1} \left[\frac{(x-1)^{n+1}}{n+1}\right]_0^{\frac{1}{2}}$$

$$= -\ln^2\left(\frac{1}{2}\right) - \sum_{n=0}^{\infty} \frac{1}{2^{n+1}(n+1)^2} + \sum_{n=0}^{\infty} \frac{1}{(n+1)^2}$$

Re-indexing both sums,

$$S = -\ln^2\left(\frac{1}{2}\right) - \underbrace{\sum_{n=1}^{\infty} \frac{1}{2^n \cdot n^2}}_{=S} + \underbrace{\sum_{n=1}^{\infty} \frac{1}{n^2}}_{=\zeta(2)}$$

6.2. SOME PROBLEMS

$$\implies S = -\ln^2\left(\frac{1}{2}\right) - S + \frac{\pi^2}{6}$$

Finally solving for S,

$$2S = -\ln^2\left(\frac{1}{2}\right) + \frac{\pi^2}{6}$$

$$S = \frac{\pi^2}{12} - \frac{\ln^2 2}{2}$$

Example 3: Define $f(n) = \int_0^1 \frac{\ln(1-x^n)}{x}dx$. Evaluate $\sum_{n=1}^{\infty} \frac{f(n)}{n}$

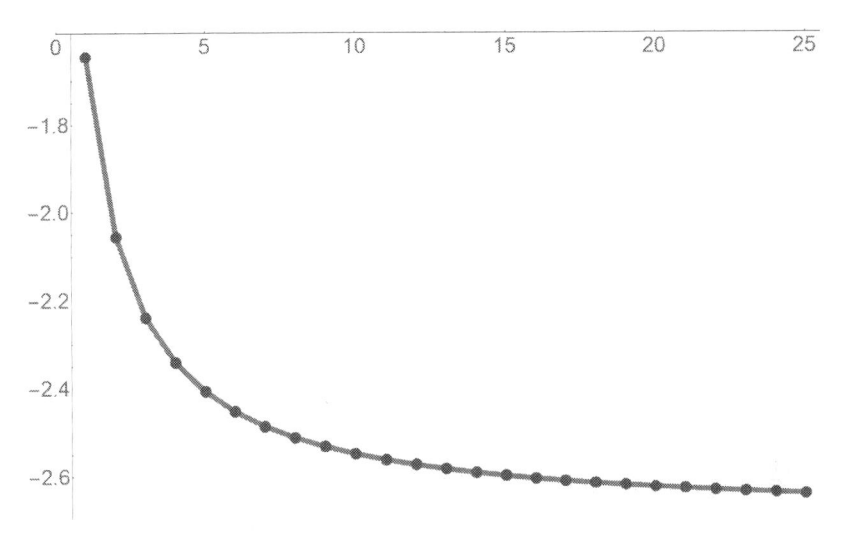

Figure 6.3: Plot of the partial sums

Solution

Consider the power series expansion of $\ln(1-x)$, (6.3):

$$\ln(1-x) = -\sum_{k=1}^{\infty} \frac{x^k}{k}$$

$$\implies f(n) = -\int_0^1 \frac{\sum_{k=1}^\infty \frac{(x^n)^k}{k}}{x} dx$$

$$= -\int_0^1 \sum_{k=1}^\infty \frac{x^{nk-1}}{k} dx$$

By the dominated convergence theorem, we can interchange summation and integration to obtain

$$f(n) = -\sum_{k=1}^\infty \int_0^1 \frac{x^{nk-1}}{k} dx$$

In evaluating the integral we obtain:

$$f(n) = -\sum_{k=1}^\infty \left[\frac{x^{nk}}{nk^2}\right]_0^1$$

$$= -\frac{1}{n}\sum_{k=1}^\infty \frac{1}{k^2}$$

$$= -\frac{\zeta(2)}{n}$$

Plugging in the value of $\zeta(2)$,

$$\therefore \int_0^1 \frac{\ln(1-x^n)}{x} dx = -\frac{\pi^2}{6n}$$

We can now evaluate our desired series:

$$S = \sum_{n=1}^\infty \frac{f(n)}{n}$$

$$= -\frac{\pi^2}{6}\sum_{n=1}^\infty \frac{1}{n^2} = -\frac{\pi^4}{36}$$

6.2. SOME PROBLEMS

Example 4: A crazy ant is standing on the origin. It begins by walking 1 unit in the $+x$ direction and then turns 60° counterclockwise and walks $\frac{1}{2}$ units in that direction. The ant then turns another 60° and walks $\frac{1}{3}$ units in that direction. The ant keeps doing this endlessly. What is the ant's final position?

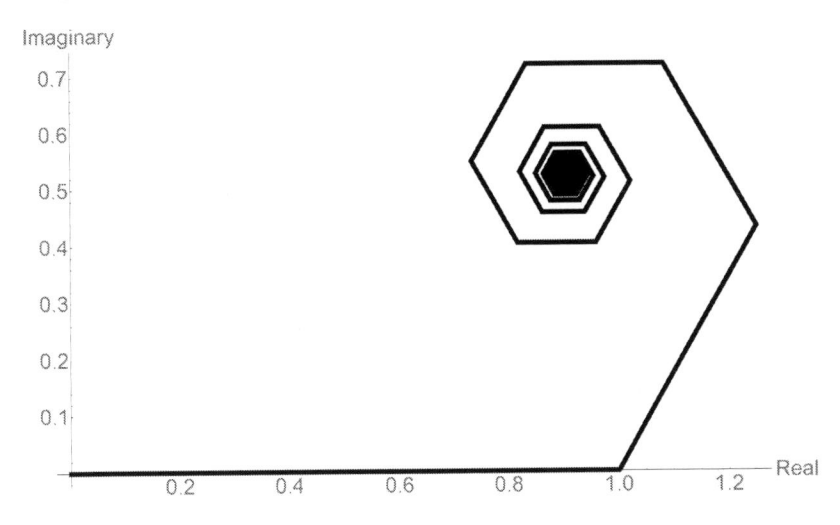

Figure 6.4: Visualization of the ant's path

Solution

Instead of using the Cartesian coordinate system, we will use the polar coordinate system to make calculations easier. This allows us to express each move the ant makes in the form

$$\frac{1}{n}[\cos\theta + i\sin\theta]$$

Where $\frac{1}{n}$ denotes the distance travelled and θ denotes the angle at which the ant is travelling.

The final position of the ant is simply the sum of the displacements. Denote the final position as P,

$$P = 1\left(\cos(0) + i\sin(0)\right) + \frac{1}{2}\left(\cos\left(\frac{\pi}{3}\right) + i\sin\left(\frac{\pi}{3}\right)\right)$$

$$+\frac{1}{3}\left(\cos\left(\frac{2\pi}{3}\right) + i\sin\left(\frac{2\pi}{3}\right)\right) + \cdots$$

Now seems like an opportune moment to use Euler's formula! (See Appendix A for more.)

$$e^{ix} = \cos x + i \sin x$$

Applying this formula to P gives

$$P = \sum_{n=1}^{\infty} \frac{e^{\frac{i(n-1)\pi}{3}}}{n}$$

This looks very similar to the power series expansion for $\ln(1-x)$. In fact, the power series expansion of $\ln(1-x)$ divided by x is

$$\frac{\ln(1-x)}{x} = -\sum_{n=1}^{\infty} \frac{x^{n-1}}{n}$$

In our desired sum, $x = e^{\frac{i\pi}{3}}$, therefore,

$$P = -\frac{\ln\left(1 - e^{\frac{i\pi}{3}}\right)}{e^{\frac{i\pi}{3}}}$$

$$= \frac{\frac{i\pi}{3}}{\frac{1+i\sqrt{3}}{2}}$$

Simplifying using the complex conjugate of the denominator gives

$$P = \frac{\pi}{2\sqrt{3}} + \frac{\pi i}{6}$$

6.2. SOME PROBLEMS

We can translate our result to the Cartesian plane and determine that our ant friend will be at:

$$P = \left(\frac{\pi}{2\sqrt{3}}, \frac{\pi}{6}\right)$$

A whole $\frac{\pi}{3} \approx 1.05$ units away!

Example 5: Evaluate $\displaystyle\sum_{n=1}^{\infty}\left(\frac{1}{4n-1} - \frac{1}{4n}\right)$

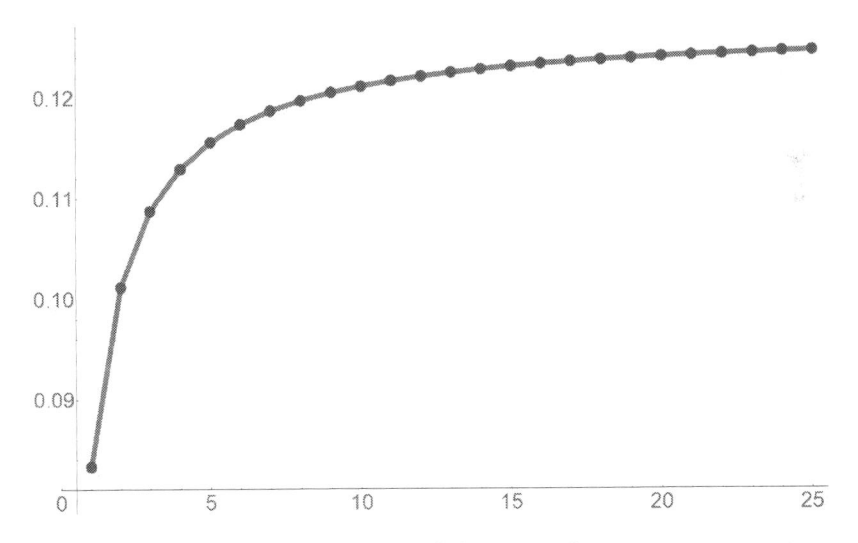

Figure 6.5: Plot of the partial sums

Solution

Although this series might first look like it is a telescoping one, it is not. One can easily see this by expanding the series,

$$S = \left(\frac{1}{4-1} - \frac{1}{4}\right) + \left(\frac{1}{4\cdot 2 - 1} - \frac{1}{4\cdot 2}\right) + \left(\frac{1}{4\cdot 3 - 1} - \frac{1}{4\cdot 3}\right) + \cdots$$

No terms cancel out. Therefore, we have to think of a different approach. Consider:

$$\frac{1}{4n-1} - \frac{1}{4n} = \int_0^1 x^{4n-2} dx - \int_0^1 x^{4n-1} dx$$

$$= \int_0^1 x^{4n-2} - x^{4n-1} dx$$

Using the above equation, we can then write our series as a sum of integrals. We can then use the dominated convergence theorem to interchange the order of summation and integration and obtain a geometric series that is easy to evaluate. We begin with

$$S = \sum_{n=1}^{\infty} \int_0^1 x^{4n-2} + x^{4n-1} dx$$

$$= \int_0^1 \sum_{n=1}^{\infty} x^{4n-2} - x^{4n-1} dx$$

$$= \int_0^1 \left(\sum_{n=1}^{\infty} \frac{(x^4)^n}{x^2} - \sum_{n=1}^{\infty} \frac{(x^4)^n}{x} \right) dx$$

Evaluating the series above,

$$S = \int_0^1 \frac{1}{x^2} \cdot \frac{x^4}{1-x^4} - \frac{1}{x} \cdot \frac{x^4}{1-x^4} dx$$

$$= \int_0^1 \frac{x^2}{1-x^4} - \frac{x^3}{1-x^4} dx$$

$$= \int_0^1 \frac{x^2(1-x)}{1-x^4} dx$$

6.2. SOME PROBLEMS

Using

$$1 - x^4 = (1-x^2)(1+x^2) = (1-x)(1+x+x^2+x^3)$$

We can write

$$S = \int_0^1 \frac{x^2}{x^3+x^2+x+1} dx$$

Using the method of partial fractions,

$$S = \frac{1}{2}\int_0^1 \frac{x-1}{x^2+1} + \frac{1}{x+1} dx$$

$$= \frac{1}{2}\left[\ln|x+1|\right]_0^1 + \frac{1}{2}\int_0^1 \frac{x-1}{x^2+1} dx$$

$$= \frac{\ln 2}{2} + \frac{1}{2}\int_0^1 \frac{x}{x^2+1} dx - \frac{1}{2}\int_0^1 \frac{1}{x^2+1} dx$$

Both integrals are easy to evaluate. Therefore,

$$S = \frac{\ln 2}{2} + \frac{\ln 2}{4} - \frac{\pi}{8}$$

$$= \frac{3\ln 2}{4} - \frac{\pi}{8}$$

6.2.1 Harmonic Numbers

> **Definition**
>
> Recall the harmonic numbers, usually denoted as H_n, from the definition of the Euler-Mascheroni constant (See note in (1.8)). They are defined as the partial sum of the harmonic series:
> $$H_n = \sum_{k=1}^{n} \frac{1}{k} \qquad (6.5)$$

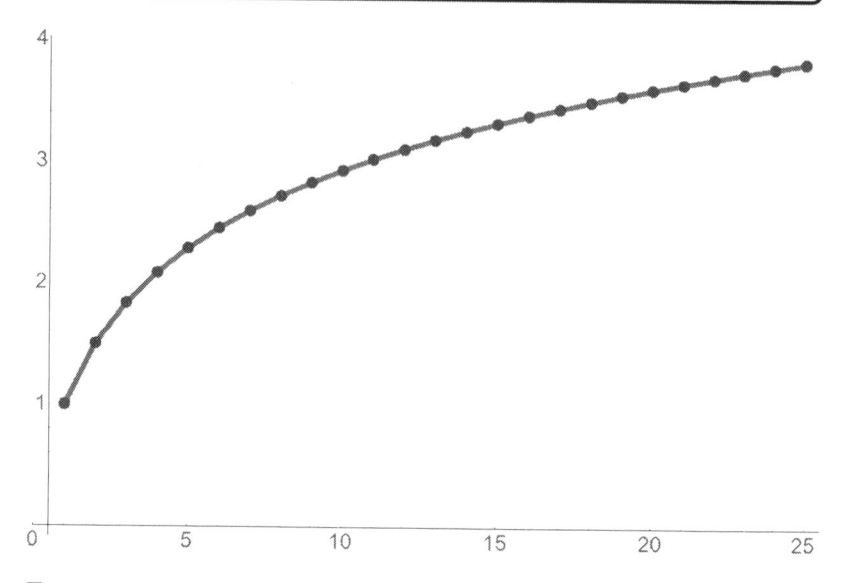

Figure 6.6: Plot of the partial sums of the harmonic series, i.e. the harmonic numbers

We can try to define an integral form of the harmonic numbers. Notice that

$$\int_0^1 x^{k-1} \mathrm{d}x = \frac{1}{k}$$

Therefore,

6.2. SOME PROBLEMS

$$H_n = \sum_{k=1}^{n} \int_0^1 x^{k-1} \, \mathrm{d}x$$

Since we have a finite sum, we can safely interchange summation and integration to obtain

$$H_n = \int_0^1 \sum_{k=1}^{n} x^{k-1} \, \mathrm{d}x$$

We can apply the geometric sum formula here

$$H_n = \int_0^1 \frac{1-x^n}{1-x} \, \mathrm{d}x \tag{6.6}$$

Indeed, this form is the well-known extension of the harmonic numbers to the complex plane other than the negative integers via analytic continuation.

Example 6: Prove that $H_n = n \sum_{k=1}^{\infty} \frac{1}{k(n+k)}$

Note that for any integer n we have

$$\lim_{x \to \infty} H_x - H_{n+x} = 0$$

We can add H_n to both sides to obtain

$$H_n = \lim_{x \to \infty} H_x - (H_{n+x} - H_n)$$

Notice that $H_{n+x} - H_n$ is simply the sum $\frac{1}{n+1} + \frac{1}{n+2} + \cdots + \frac{1}{n+x}$. Therefore,

$$H_n = \lim_{x \to \infty} \sum_{k=1}^{x} \frac{1}{k} - \sum_{k=1}^{x} \frac{1}{n+k} \qquad (6.7)$$

$$= \lim_{x \to \infty} \sum_{k=1}^{x} \left(\frac{1}{k} - \frac{1}{n+k} \right)$$

$$= \lim_{x \to \infty} \sum_{k=1}^{x} \frac{n}{k(n+k)}$$

$$\therefore H_n = n \sum_{k=1}^{\infty} \frac{1}{k(n+k)} \qquad (6.8)$$

We can derive more identities about harmonic numbers.

Theorem

The generating function for the harmonic numbers is given by:
$$\sum_{n=1}^{\infty} H_n x^n = \frac{\ln(1-x)}{x-1} \qquad (6.9)$$

For any $|x| < 1$.

Proof. Consider the definition of the harmonic numbers:

$$H_n = H_{n-1} + \frac{1}{n}$$

$$\therefore H_n - H_{n-1} = \frac{1}{n}$$

Multiplying both sides by x^n,

$$H_n x^n - H_{n-1} x^n = \frac{x^n}{n}$$

6.2. SOME PROBLEMS

Summing both sides from $n = 1$ to ∞,

$$\sum_{n=1}^{\infty} H_n x^n - \sum_{n=1}^{\infty} H_{n-1} x^n = \sum_{n=1}^{\infty} \frac{x^n}{n}$$

Notice that the RHS is $-\ln(1-x)$. Therefore,

$$\sum_{n=1}^{\infty} H_n x^n - \sum_{n=1}^{\infty} H_{n-1} x^n = -\ln(1-x)$$

Factoring out an x from the second sum,

$$\sum_{n=1}^{\infty} H_n x^n - x \sum_{n=1}^{\infty} H_{n-1} x^{n-1} = -\ln(1-x)$$

Since $H_0 = 0$,

$$\sum_{n=1}^{\infty} H_n x^n = \sum_{n=1}^{\infty} H_{n-1} x^{n-1}$$

And,

$$(1-x) \sum_{n=1}^{\infty} H_n x^n = -\ln(1-x)$$

Therefore,

$$\sum_{n=1}^{\infty} H_n x^n = \frac{\ln(1-x)}{x-1}$$

□

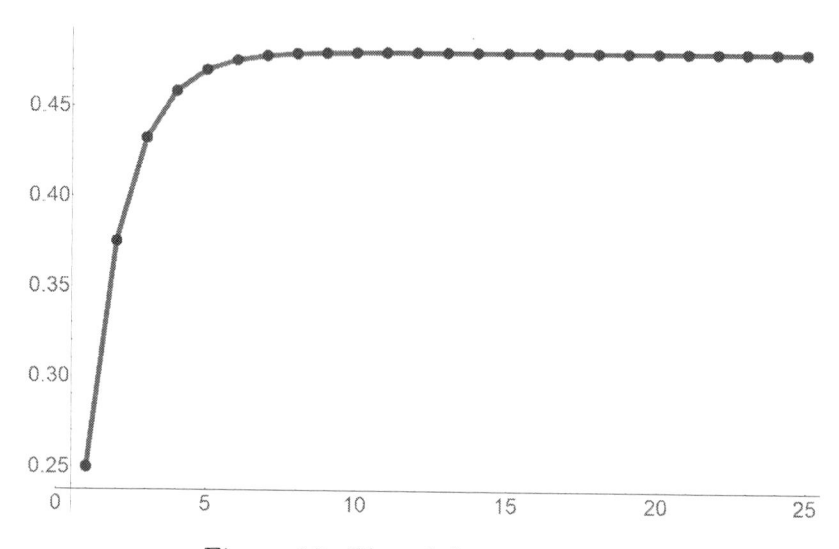

Example 7: Evaluate $\displaystyle\sum_{n=1}^{\infty}\frac{H_n}{2^n(n+1)}$

Figure 6.7: Plot of the partial sums

Solution

The series converges by the ratio test. Consider (6.9)

$$\sum_{n=1}^{\infty} H_n x^n = \frac{\ln(1-x)}{x-1}$$

Integrating both sides from $x=0$ to $x=\frac{1}{2}$,

$$\int_0^{\frac{1}{2}} \sum_{n=1}^{\infty} H_n x^n \,\mathrm{d}x = \int_0^{\frac{1}{2}} \frac{\ln(1-x)}{x-1}\,\mathrm{d}x$$

We can now use the dominated convergence theorem to interchange summation and integration,

6.2. SOME PROBLEMS

$$\sum_{n=1}^{\infty} \int_0^{\frac{1}{2}} H_n x^n \mathrm{d}x = \int_0^{\frac{1}{2}} \frac{\ln(1-x)}{x-1} \mathrm{d}x$$

$$\sum_{n=1}^{\infty} H_n \left[\frac{x^{n+1}}{n+1}\right]_0^{\frac{1}{2}} = \int_0^{\frac{1}{2}} \frac{\ln(1-x)}{x-1} \mathrm{d}x$$

$$\sum_{n=1}^{\infty} \frac{H_n}{2^{n+1}(n+1)} = -\underbrace{\int_0^{\frac{1}{2}} \frac{\ln(1-x)}{1-x} \mathrm{d}x}_{I} \qquad (6.10)$$

For the integral on the RHS, consider IBP with $u = \ln(1-x)$, $dv = \frac{dx}{1-x}$:

$$-I = \int_0^{\frac{1}{2}} \frac{\ln(1-x)}{1-x} \mathrm{d}x = \left[\ln^2(1-x)\right]_0^{\frac{1}{2}} - \underbrace{\int_0^{\frac{1}{2}} \frac{\ln(1-x)}{1-x} \mathrm{d}x}_{-I}$$

Hence,

$$-2I = \left[\ln^2(1-x)\right]_0^{\frac{1}{2}}$$

$$\therefore I = \frac{\ln^2 2}{2}$$

Plugging I back into (6.10) gives:

$$\sum_{n=1}^{\infty} \frac{H_n}{2^{n+1}(n+1)} = \frac{\ln^2 2}{2}$$

$$\therefore \sum_{n=1}^{\infty} \frac{H_n}{2^n(n+1)} = \ln^2 2$$

Example 8: Evaluate $S_n = \sum_{k=1}^{n} \frac{(-1)^k \binom{n}{k}}{k}$

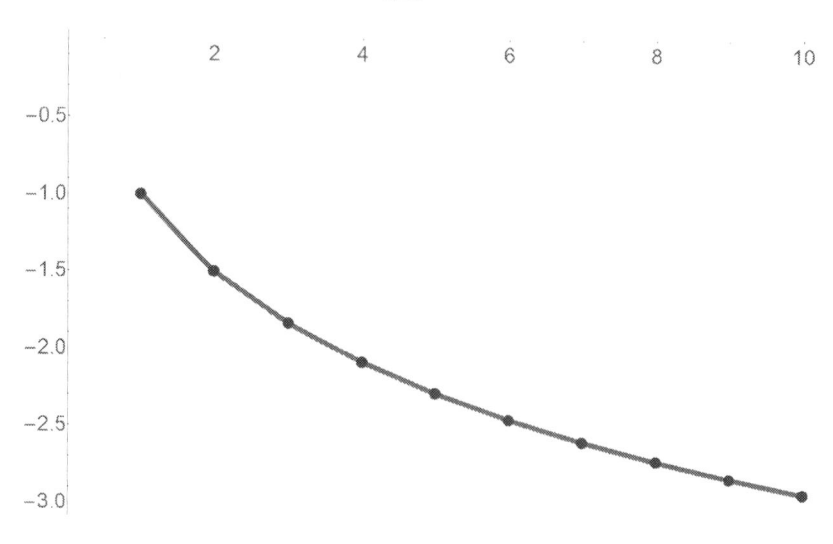

Figure 6.8: Plot of S_n values

Solution

Recall the statement of the binomial theorem:

$$(x+y)^n = \sum_{k=0}^{n} \binom{n}{k} x^{n-k} y^k$$

We therefore have

$$(1-x)^n = \sum_{k=0}^{n} \binom{n}{k} (-x)^k$$

$$\therefore \sum_{k=1}^{n} \binom{n}{k} (-x)^k = (1-x)^n - 1$$

Dividing by x then gives

6.2. SOME PROBLEMS

$$\sum_{k=1}^{n} \binom{n}{k}(-1)^k x^{k-1} = \frac{(1-x)^n - 1}{x}$$

Integrating both sides from $x = 0$ to $x = 1$,

$$\int_0^1 \sum_{k=1}^{n} \binom{n}{k}(-1)^k x^{k-1} dx = \int_0^1 \frac{(1-x)^n - 1}{x} dx$$

Since our sum on the LHS is a finite sum, we can safely interchange summation and integration

$$\sum_{k=1}^{n}(-1)^k \binom{n}{k} \int_0^1 x^{k-1} dx = \int_0^1 \frac{(1-x)^n - 1}{x} dx$$

$$\sum_{k=1}^{n} \frac{(-1)^k \binom{n}{k}}{k} = \int_0^1 \frac{(1-x)^n - 1}{x} dx$$

Using (2.1), we can write

$$\sum_{k=1}^{n} \frac{(-1)^k \binom{n}{k}}{k} = \int_0^1 \frac{x^n - 1}{1-x} dx$$

By (6.6), this is simply $-H_n$! We can then express our sum as

$$\sum_{k=1}^{n} \frac{(-1)^k \binom{n}{k}}{k} = -H_n = -\sum_{k=1}^{n} \frac{1}{k}$$

This could also be a nice proof for the divergence of the harmonic series, as the series on the left diverges as $n \to \infty$.

Example 9: Evaluate $\sum_{n=1}^{\infty} \dfrac{H_n}{n^3}$

Solution

Consider the result in (6.8):

$$H_n = \sum_{k=1}^{\infty} \frac{n}{k(k+n)}$$

We can then express our sum as

$$S = \sum_{n=1}^{\infty} \frac{H_n}{n^3}$$

$$= \sum_{n=1}^{\infty} \frac{1}{n^3} \sum_{k=1}^{\infty} \frac{n}{k(k+n)}$$

$$= \sum_{n=1}^{\infty} \sum_{k=1}^{\infty} \frac{1}{n^2 k(k+n)}$$

By symmetry,

$$\sum_{n=1}^{\infty} \sum_{k=1}^{\infty} \frac{1}{n^2 k(k+n)} = \sum_{n=1}^{\infty} \sum_{k=1}^{\infty} \frac{1}{nk^2(k+n)}$$

Thus,

$$S = \frac{1}{2} \left[\sum_{n=1}^{\infty} \sum_{k=1}^{\infty} \frac{1}{n^2 k(k+n)} + \sum_{n=1}^{\infty} \sum_{k=1}^{\infty} \frac{1}{nk^2(k+n)} \right]$$

6.3. EXERCISE PROBLEMS

$$= \frac{1}{2}\sum_{n=1}^{\infty}\sum_{k=1}^{\infty}\left(\frac{1}{n^2k(k+n)} + \frac{1}{nk^2(k+n)}\right)$$

Factoring out a $\frac{1}{nk(k+n)}$ gives:

$$S = \frac{1}{2}\sum_{n=1}^{\infty}\sum_{k=1}^{\infty}\frac{1}{nk(k+n)}\left(\frac{1}{n} + \frac{1}{k}\right)$$

$$= \frac{1}{2}\sum_{n=1}^{\infty}\sum_{k=1}^{\infty}\frac{1}{nk\cancel{(k+n)}}\left(\frac{\cancel{n+k}}{nk}\right)$$

$$= \frac{1}{2}\sum_{n=1}^{\infty}\sum_{k=1}^{\infty}\frac{1}{n^2k^2}$$

$$= \frac{1}{2}\sum_{n=1}^{\infty}\frac{1}{n^2}\sum_{k=1}^{\infty}\frac{1}{k^2}$$

$$= \frac{1}{2}\sum_{n=1}^{\infty}\frac{\zeta(2)}{n^2}$$

$$\therefore \sum_{n=1}^{\infty}\frac{H_n}{n^3} = \frac{(\zeta(2))^2}{2} = \frac{\pi^4}{72} \qquad (6.11)$$

6.3 Exercise Problems

1) Evaluate $\sum_{n=1}^{\infty}\frac{n}{(2n+1)!}$ (Hint: Use the Taylor series of e^x)

2) Find $\sum_{k=1}^{\infty} \frac{\sin k°}{k}$ (Hint: Convert into radians and use Euler's formula)

3) Evaluate $\sum_{n=0}^{\infty} \frac{2^n}{(2n+1)\binom{2n}{n}}$

(Hint: Use the identity $\int_0^{\pi/2} \sin^{2k+1} x \, dx = \frac{2^{2k} k!^2}{(2k+1)!}$)

4) Find the value of $\sum_{n=1}^{\infty} \frac{(-1)^n H_n}{n}$

5) Evaluate $\sum_{n=0}^{k} \frac{(-1)^n}{\binom{k}{n}}$ for even k.

Chapter 7

Series and Integrals

7.1 Introduction

In this chapter, we will apply both the standard integration techniques given in chapter 2 as well as the skills we developed with series in the last three chapters. This will serve as our introduction to advanced integrals, as most of these integrals are impossible or very hard to solve using the standard techniques of $u-$substitution, IBP, etc, alone. It will also allow us to have an easy transition to the much more involved chapters dealing with special functions!

7.2 Some Problems

As always, we will begin with an easy problem.

Example 1: Evaluate $\displaystyle\int_0^\infty \frac{\ln x}{1-x^2}\,\mathrm{d}x$

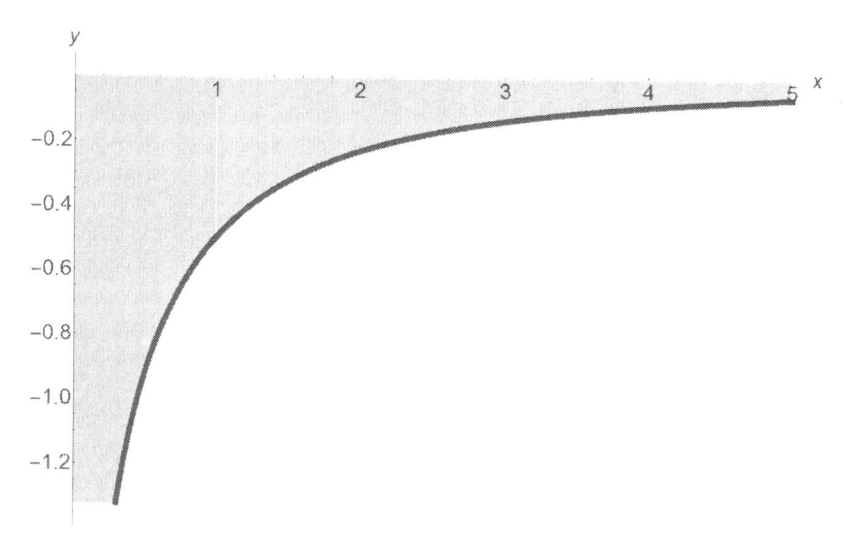

Figure 7.1: Graph of $y = \frac{\ln x}{1-x^2}$

7.2. SOME PROBLEMS

Solution
Splitting the integral into two parts,

$$I = \int_0^1 \frac{\ln x}{1-x^2} dx + \underline{\int_1^\infty \frac{\ln x}{1-x^2} dx}$$

Letting $x = \frac{1}{u}$, $dx = -\frac{1}{u^2} du$ in the underlined integral,

$$\int_1^\infty \frac{\ln x}{1-x^2} dx = -\int_1^0 \frac{\ln\left(\frac{1}{u}\right)}{u^2 \left(1 - \frac{1}{u^2}\right)} du$$

$$= \int_0^1 \frac{\ln u}{1-u^2} du$$

Therefore,

$$I = 2\int_0^1 \frac{\ln x}{1-x^2} dx$$

By the power series expansion of $\frac{1}{1-x}$, we can write

$$I = 2\int_0^1 \sum_{n=0}^\infty x^{2n} \ln x \, dx$$

$$= 2\sum_{n=0}^\infty \int_0^\infty x^{2n} \ln x \, dx$$

Where we used the dominated convergence theorem in the last step. By IBP with $u = \ln x$, $dv = x^{2n} dx$, we have

$$I = \sum_{n=0}^\infty \left(\left[\frac{x^{2n+1} \ln x}{2n+1}\right]_0^1 - \frac{1}{2n+1} \int_0^1 x^{2n} dx \right)$$

$$= -\sum_{n=0}^{\infty} \frac{1}{(2n+1)^2}$$

This series can be easily evaluated. Consider

$$\zeta(2) = \sum_{n=1}^{\infty} \frac{1}{n^2}$$

$$= \sum_{n=1}^{\infty} \frac{1}{(2n)^2} + \sum_{n=1}^{\infty} \frac{1}{(2n-1)^2}$$

$$= \frac{1}{4} \sum_{n=1}^{\infty} \frac{1}{n^2} + \sum_{n=1}^{\infty} \frac{1}{(2n-1)^2}$$

$$\therefore \frac{3}{4}\zeta(2) = \frac{\pi^2}{8} = \sum_{n=1}^{\infty} \frac{1}{(2n-1)^2}$$

We finally have

$$I = -2 \times \frac{\pi^2}{8}$$

$$= -\frac{\pi^2}{4}$$

Example 2: Evaluate $\int_0^{\frac{\pi}{2}} \ln(\cos x + \sin x) \mathrm{d}x$

Solution

Recall that

$$\sin(x+y) = \sin x \cos y + \sin y \cos x$$

Thus,

7.2. SOME PROBLEMS

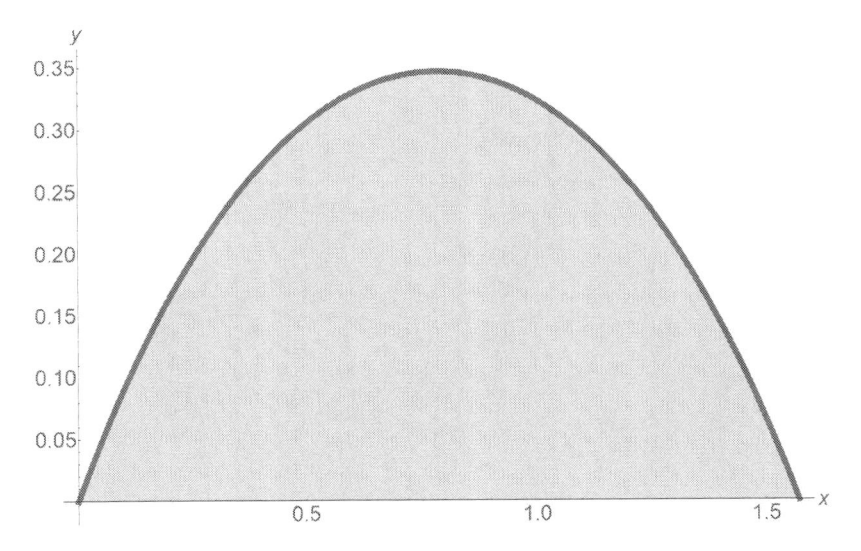

Figure 7.2: Graph of $y = \ln(\cos x + \sin x)$

$$\sin\left(x + \frac{\pi}{4}\right) = \cos\left(\frac{\pi}{4}\right)\sin x + \sin\left(\frac{\pi}{4}\right)\cos x$$

$$= \frac{\sqrt{2}}{2}(\sin x + \cos x)$$

$$\implies \sin x + \cos x = \sqrt{2}\sin\left(x + \frac{\pi}{4}\right)$$

Hence,

$$I = \int_0^{\frac{\pi}{2}} \ln\left(\sqrt{2}\sin\left(x + \frac{\pi}{4}\right)\right) dx$$

Splitting the integral in half,

$$I = \int_0^{\frac{\pi}{4}} \ln\left(\sqrt{2}\sin\left(x+\frac{\pi}{4}\right)\right)dx + \int_{\frac{\pi}{4}}^{\frac{\pi}{2}} \ln\left(\sqrt{2}\sin\left(x+\frac{\pi}{4}\right)\right)dx$$

$$\underbrace{\phantom{\int_0^{\frac{\pi}{4}} \ln\left(\sqrt{2}\sin\left(x+\frac{\pi}{4}\right)\right)dx}}_{(1)} \quad \underbrace{\phantom{\int_{\frac{\pi}{4}}^{\frac{\pi}{2}} \ln\left(\sqrt{2}\sin\left(x+\frac{\pi}{4}\right)\right)dx}}_{(2)}$$

Applying our integral reflection identity to (1) and substituting $x \to x + \frac{\pi}{4}$ in (2) gives

$$I = \int_0^{\frac{\pi}{4}} \ln\left(\sqrt{2}\sin\left(\frac{\pi}{2}-x\right)\right)dx + \int_0^{\frac{\pi}{4}} \ln\left(\sqrt{2}\sin\left(x+\frac{\pi}{2}\right)\right)dx$$

Notice that $\sin\left(\frac{\pi}{2}-x\right) = \sin\left(x+\frac{\pi}{2}\right) = \cos x$.

$$\implies I = 2\int_0^{\frac{\pi}{4}} \ln\left(\sqrt{2}\cos x\right)dx$$

$$= \frac{\pi \ln 2}{4} + 2\int_0^{\frac{\pi}{4}} \ln(\cos x)\,dx$$

Now, define two integrals:

$$I_1 = \int_0^{\frac{\pi}{4}} \ln(\cos x)\,dx$$

$$I_2 = \int_0^{\frac{\pi}{4}} \ln(\sin x)\,dx$$

A creative strategy to crack our desired integral, I_1, is to set up a system of linear equations with I_1 and I_2 being the variables. We will first add I_1 and I_2

$$I_1 + I_2 = \int_0^{\frac{\pi}{4}} \ln(\cos x \sin x)\,dx$$

7.2. SOME PROBLEMS

$$= \int_0^{\frac{\pi}{4}} \ln\left(\frac{1}{2}\sin 2x\right) dx$$

$$= \int_0^{\frac{\pi}{4}} \ln(\sin 2x) \, dx - \int_0^{\frac{\pi}{4}} \ln 2 \, dx$$

Notice that both integrands above are symmetric around $x = \frac{\pi}{4}$ on the interval $\left[0, \frac{\pi}{2}\right]$. Thus,

$$I_1 + I_2 = \frac{1}{2}\left[\int_0^{\frac{\pi}{2}} \ln(\sin 2x) \, dx - \int_0^{\frac{\pi}{2}} \ln 2 \, dx\right]$$

The bracketed expression has been evaluated previously (See (2.2)). Therefore,

$$I_1 + I_2 = -\frac{\pi}{2}\ln 2$$

Moving on, we can compute the difference between I_2 and I_1,

$$I_2 - I_1 = \int_0^{\frac{\pi}{4}} \ln(\tan x) \, dx$$

Substituting $u = \tan x$, $dx = \frac{du}{1+u^2}$ then gives

$$I_2 - I_1 = \int_0^1 \frac{\ln u}{1+u^2} du$$

$$= \int_0^1 \sum_{n=0}^{\infty} (-1)^n u^{2n} \ln u \, du$$

Where we applied the series expansion of $\frac{1}{1+u^2}$ in the last step. By the dominated convergence theorem, we can switch the order of summation and integration to obtain

$$I_2 - I_1 = \sum_{n=0}^{\infty} (-1)^n \int_0^1 u^{2n} \ln u \, du$$

Using IBP,

$$I_2 - I_1 = -\sum_{n=0}^{\infty} \frac{(-1)^n}{(2n+1)^2}$$
$$= -G$$

Where G is **Catalan's constant**.

> **Definition**
>
> Catalan's constant, often denoted as G, is named after the French and Belgian mathematician Eugène Charles Catalan. It was originally found in the field of combinatorics, but later found use in analysis and even topology[a]. It is mainly defined as the infinite series
>
> $$G = \sum_{n=0}^{\infty} \frac{(-1)^n}{(2n+1)^2} \qquad (7.1)$$
>
> ---
> [a] Agol, Ian (2010). *The minimal volume orientable hyperbolic 2-cusped 3-manifolds*. Proceedings of the American Mathematical Society, 138 (10): 3723–3732, arXiv:0804.0043, doi:10.1090/S0002-9939-10-10364-5, MR 2661571.

> **Open Problem**
>
> It is not known whether Catalan's constant is irrational, let alone transcendental[a]. Proving the irrationality or transcendentality of this constant would be a major result.
>
> ---
> [a] Nesterenko, Yu. V. (January 2016). *On Catalan's constant*. Proceedings of the Steklov Institute of Mathematics, 292 (1): 153–170, doi:10.1134/s0081543816010107.

7.2. SOME PROBLEMS

To summarize,
$$I_1 + I_2 = \frac{-\pi \ln 2}{2}$$
$$I_2 - I_1 = -G$$

Therefore,
$$I_1 = \int_0^{\frac{\pi}{4}} \ln \cos x \, dx = \frac{1}{2}\left(G - \frac{\pi \ln 2}{2}\right) \quad (7.2)$$

$$I_2 = \int_0^{\frac{\pi}{4}} \ln \sin x \, dx = -\frac{1}{2}\left(G + \frac{\pi \ln 2}{2}\right) \quad (7.3)$$

We finally have
$$I = \frac{\pi \ln 2}{4} + 2I_1$$
$$= G - \frac{\pi \ln 2}{4}$$

Example 3: Define $f(\alpha) = \int_0^{\alpha} \ln x \ln(\alpha - x) \, dx$. Find when the minimum of $f(\alpha)$ occurs.

Solution

The substitution $x = \alpha u$, $dx = \alpha du$ transforms our integral into

$$f(\alpha) = \alpha \int_0^1 \ln(\alpha u) \ln\bigl(\alpha(1-u)\bigr) \, du$$
$$= \alpha \int_0^1 [\ln \alpha + \ln(1-u)][\ln \alpha + \ln u] \, du$$

Expanding the integrand,

224   CHAPTER 7. SERIES AND INTEGRALS

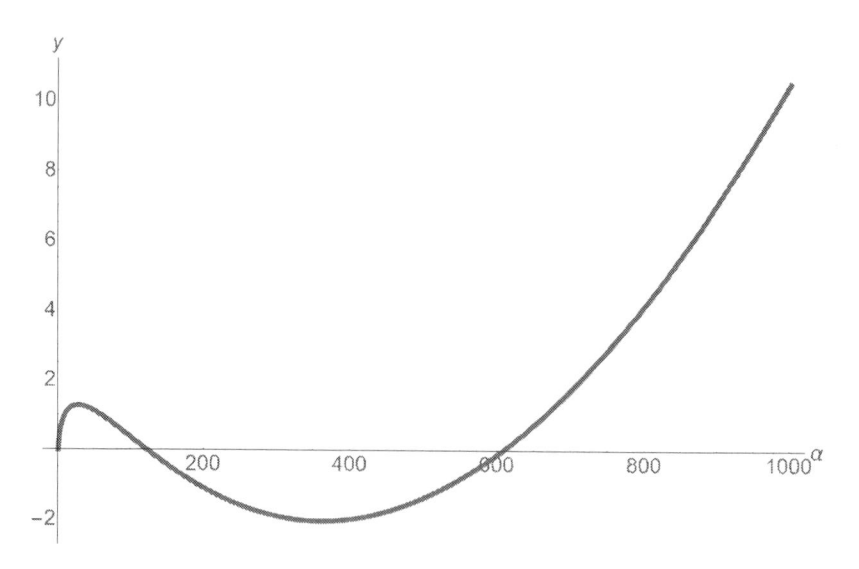

Figure 7.3: Graph of $y = f(\alpha)$

$$f(\alpha) = \alpha \int_0^1 \ln^2 \alpha + \ln \alpha \ln u + \ln \alpha \ln(1-u) + \ln(1-u) \ln u \ du$$

$$= \alpha \left[\int_0^1 \ln^2 \alpha \ du + \int_0^1 \ln \alpha \ln u \ du \right.$$
$$\left. + \int_0^1 \ln \alpha \ln(1-u) du + \int_0^1 \ln u \ln(1-u) \ du \right]$$

The first three integrals are trivial to evaluate. For the last integral, consider the power series expansion of $\ln(1-u)$:

$$\ln(1-u) = -\sum_{n=1}^{\infty} \frac{u^n}{n}$$

$$\therefore \mathcal{I} = \int_0^1 \ln u \ln(1-u) du = -\int_0^1 \sum_{n=1}^{\infty} \frac{u^n \ln u}{n}$$

7.2. SOME PROBLEMS

By the dominated convergence theorem, we can interchange summation and integration to get:

$$\mathcal{I} = -\sum_{n=1}^{\infty} \frac{1}{n} \int_0^1 u^n \ln u \, du$$

$$= \sum_{n=1}^{\infty} \frac{1}{n(n+1)^2}$$

Notice that we can break up this sum into:

$$\mathcal{I} = -\sum_{n=1}^{\infty} \left[\frac{1}{n} - \frac{1}{n+1} - \frac{1}{(n+1)^2} \right]$$

$$= \underbrace{\sum_{n=1}^{\infty} \left(\frac{1}{n} - \frac{1}{n+1} \right)}_{\text{Telescoping}} - \sum_{n=1}^{\infty} \frac{1}{(n+1)^2}$$

$$= 2 - \frac{\pi^2}{6}$$

We then have,

$$f(\alpha) = \alpha \left[\ln^2 \alpha + 2 - 2\ln \alpha - \frac{\pi^2}{6} \right]$$

Therefore,

$$f'(\alpha) = \ln^2 \alpha - \frac{\pi^2}{6}$$

$$f''(\alpha) = \frac{2 \ln \alpha}{\alpha}$$

It is now easy to see that the minimum is at $\alpha = e^{\frac{\pi}{\sqrt{6}}}$. Utilizing our expression for $f(\alpha)$ we can compute the minimum as:

$$f\left(e^{\frac{\pi}{\sqrt{6}}}\right) = e^{\frac{\pi}{\sqrt{6}}}\left[\ln^2\left(e^{\frac{\pi}{\sqrt{6}}}\right) + 2 - 2\ln\left(e^{\frac{\pi}{\sqrt{6}}}\right) - \frac{\pi^2}{6}\right]$$

$$= e^{\frac{\pi}{\sqrt{6}}}\left[\frac{\pi^2}{6} + 2 - \frac{2\pi}{\sqrt{6}} - \frac{\pi^2}{6}\right]$$

$$= 2e^{\frac{\pi}{\sqrt{6}}}\left(1 - \frac{\pi}{\sqrt{6}}\right) \approx -2.038$$

Example 4: Evaluate $\int_0^{\frac{\pi}{2}} \tan x \ln \sin x \, dx$

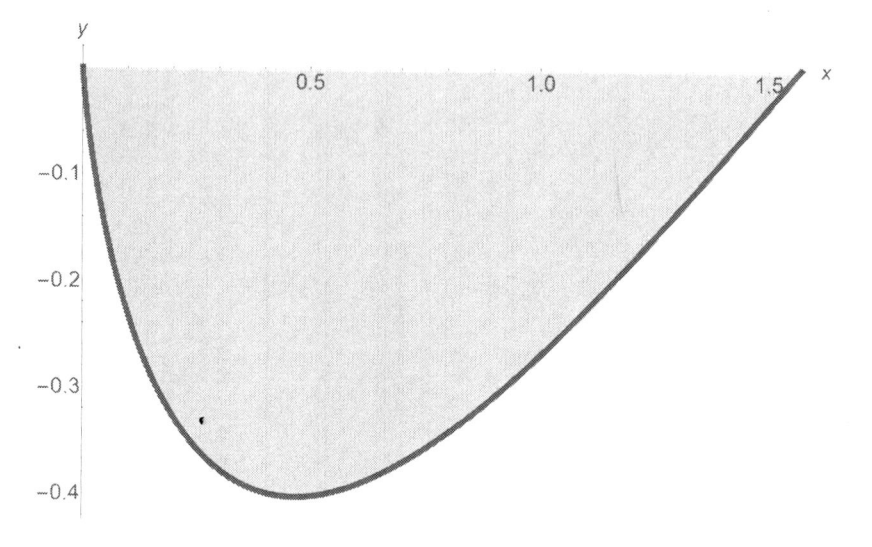

Figure 7.4: Graph of $y = \tan x \ln \sin x$

Solution

Consider rewriting the integral as:

$$I = \int_0^{\frac{\pi}{2}} \frac{\sin x}{\cos x} \ln \sin x \, dx$$

$$= \int_0^{\frac{\pi}{2}} \frac{\sin x}{\cos x} \ln\left(\sqrt{1 - \cos^2 x}\right) dx$$

7.2. SOME PROBLEMS

Substituting $u = \cos x$, $du = -\sin x \, dx$ gives

$$I = \frac{1}{2} \int_0^1 \frac{\ln(1-u^2)}{u} du$$

Now, using the power series expansion of $\ln(1-x)$, we can write

$$\ln(1-u^2) = -\sum_{n=1}^{\infty} \frac{u^{2n}}{n}$$

Therefore,

$$I = -\frac{1}{2} \int_0^1 \sum_{n=1}^{\infty} \frac{u^{2n-1}}{n} du$$

By the dominated convergence theorem, we can interchange summation and integration.

$$I = -\frac{1}{2} \sum_{n=1}^{\infty} \frac{1}{n} \int_0^1 u^{2n-1} du$$

$$= -\frac{1}{2} \sum_{n=1}^{\infty} \frac{1}{n} \left[\frac{u^{2n}}{2n} \right]_0^1$$

$$= -\frac{1}{4} \sum_{n=1}^{\infty} \frac{1}{n^2}$$

We can plug in $\sum_{n=1}^{\infty} \frac{1}{n^2} = \frac{\pi^2}{6}$ in the above expression,

$$\therefore \int_0^{\frac{\pi}{2}} \tan x \ln \sin x \, dx = -\frac{\pi^2}{24}$$

For a proof of $\sum_{n=1}^{\infty} \frac{1}{n^2} = \frac{\pi^2}{6}$, see (12.2).

Example 5: Evaluate $\displaystyle\int_0^1 \frac{\ln x \ln^2(1-x)}{x}\,dx$

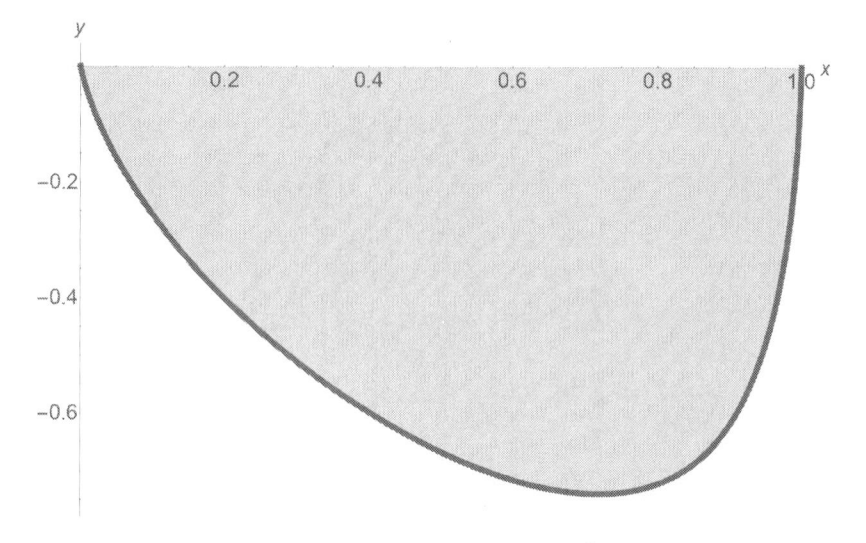

Figure 7.5: Graph of $y = \frac{\ln x \ln^2(1-x)}{x}$

Solution

By (2.1) we can express our integral as

$$I = \int_0^1 \frac{\ln(1-x)\ln^2 x}{1-x}\,dx$$

We can then use the generating function for the harmonic numbers, (6.9), to get

$$I = -\int_0^1 \sum_{n=1}^{\infty} H_n x^n \ln^2 x \, dx$$

7.2. SOME PROBLEMS

By the dominated convergence theorem, we can interchange summation and integration to obtain

$$I = -\sum_{n=1}^{\infty} H_n \int_0^1 x^n \ln^2 x \, dx$$

Now, consider the integral

$$f(\alpha) = \int_0^1 x^\alpha dx = \frac{1}{\alpha+1}$$

For $\alpha > -1$. Differentiating k times under the integral sign gives

$$\int_0^1 x^\alpha \ln^k x \, dx = \frac{(-1)^k k!}{(\alpha+1)^{k+1}}$$

Then,

$$\int_0^1 x^n \ln^2 x \, dx = \frac{(-1)^2 \cdot 2!}{(n+1)^3}$$

And

$$I = -2 \sum_{n=1}^{\infty} \frac{H_n}{(n+1)^3}$$

$$= -2 \sum_{n=1}^{\infty} \frac{1}{(n+1)^3} \left(H_{n+1} - \frac{1}{n+1} \right)$$

$$= 2 \left[\sum_{n=1}^{\infty} \frac{1}{(n+1)^4} - \sum_{n=1}^{\infty} \frac{H_{n+1}}{(n+1)^3} \right]$$

$$= 2\left[\sum_{n=1}^{\infty}\frac{1}{n^4} - \sum_{n=1}^{\infty}\frac{H_n}{n^3}\right]$$

$$\therefore I = 2\zeta(4) - 2\sum_{n=1}^{\infty}\frac{H_n}{n^3}$$

We can plug in our result for the latter sum from (6.11).

$$\implies I = 2\zeta(4) - \frac{2 \cdot \pi^4}{72}$$

Using $\zeta(4) = \frac{\pi^4}{90}$ (See proof in (12.3)) we have:

$$\int_0^1 \frac{\ln x \ln^2(1-x)}{x}\,dx = -\frac{\pi^4}{180}$$

7.3 Exercise Problems

1) Evaluate $\displaystyle\int_0^1 \ln x \ln(1-x)\,dx$

2) Find the value of $\displaystyle\int_0^\infty \frac{\sin^{2n+1} x}{x}\,dx$ for $n \in \mathbb{N}$.

3) Evaluate $\displaystyle\int_0^1 \ln\left(\frac{1+x}{1-x}\right)\frac{dx}{x\sqrt{1-x^2}}$

4) Find $\displaystyle\int_0^{\pi/4} \ln\tan x\,dx$

7.3. EXERCISE PROBLEMS

5) Evaluate $\int_0^\infty \ln^2 \tanh x \, dx$

6) Find the value of $\int_0^1 \dfrac{x \ln^2 x}{1 - x^4} dx$

7) Evaluate $\displaystyle\lim_{n \to \infty} \int_0^1 \dfrac{x^n - x^{2n}}{1 - x} dx$

8)

> **Challenge Problem**
>
> Evaluate
> $$\int_{-1}^1 \arctan x \, \arcsin x \, dx$$
> Hint: Use
> $$\arctan x = \int_0^1 \dfrac{x}{1 + x^2 y^2} dy$$

Chapter 8

Fractional Part Integrals

8.1 Introduction

In this chapter, we will be dealing with integrals that contain the **fractional part function**.

> **Definition**
>
> The fractional part function, usually denoted as $\{\cdot\}$ is defined as follows:
>
> $$\{x\} = x - \lfloor x \rfloor \quad x \geq 0 \qquad (8.1)$$
>
> Where $\lfloor \cdot \rfloor$ denotes the **floor function** (See below for definition). For example, $\{1.5\} = .5$. The fractional part function is defined for negative x, but the definitions vary.

> **Definition**
>
> The floor function, usually denoted as $\lfloor \cdot \rfloor$, is defined as the greatest integer less than or equal to x. Similarly, the ceiling function, usually denoted $\lceil x \rceil$, is defined as the least integer greater than or equal to x. Equivalently,
>
> $$\lfloor x \rfloor = \max\{n \in \mathbb{Z} \mid n \leq x\} \qquad (8.2)$$
>
> $$\lceil x \rceil = \min\{n \in \mathbb{Z} \mid n \geq x\} \qquad (8.3)$$

The results that these integrals give rise to are mind-blowing! The essence in solving problems like the ones in this chapter is converting them to infinite sums of consecutive integrals which we can manage. Although these integrals are not a primary area of research, they have a chapter dedicated to them since they develop the skill of using series to evaluate integrals.

8.2 Some Problems

We begin with the most basic of fractional part integrals:

Example 1: Evaluate $\int_0^n \{x\} \, dx$ for $n \in \mathbb{N}$.

First, we will take a look at the graph of $y = \{x\}$.

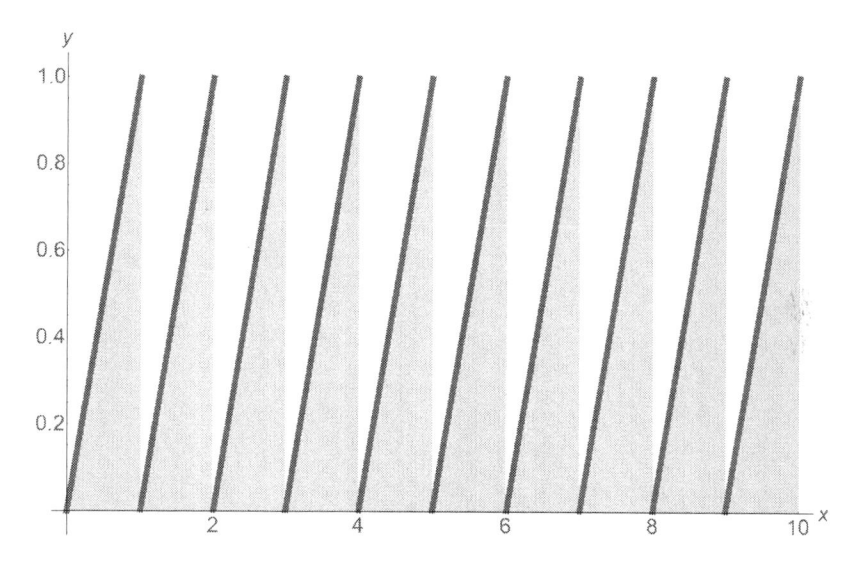

Figure 8.1: Case when $n = 10$

Notice that each line is simply $y = x$ shifted along the x axis. This is rather obvious from the definition of the fractional part function. Consider some $x \in [k, k+1)$, it is easy to see that

$$\{x\} = x - \lfloor x \rfloor = x - k$$

Therefore, the integral in our example is equivalent to a sum of integrals with intervals $[k, k+1)$ from $k = 0$ to $k = n - 1$. Translating these words into math we have

$$I = \int_0^n \{x\} \, dx = \sum_{k=0}^{n-1} \int_k^{k+1} (x - k) \, dx$$

$$= \frac{1}{2}\sum_{k=0}^{n-1} 1 = \frac{n}{2}$$

Notice the strategy in solving these integrals. Since the fractional part, floor, and ceiling functions are piece-wise functions, we can think about their definite integrals (or the area under their curves) as a sum of consecutive areas. In other words, the principal aim when cracking these fractional part integrals is to transform them to sums of integrals without fractional parts.

Now that we know how to approach such types of integrals, let us delve a little deeper.

Example 2: Evaluate $\int_0^n \{x\}\lfloor x \rfloor \, dx$ for $n \in \mathbb{N}$.

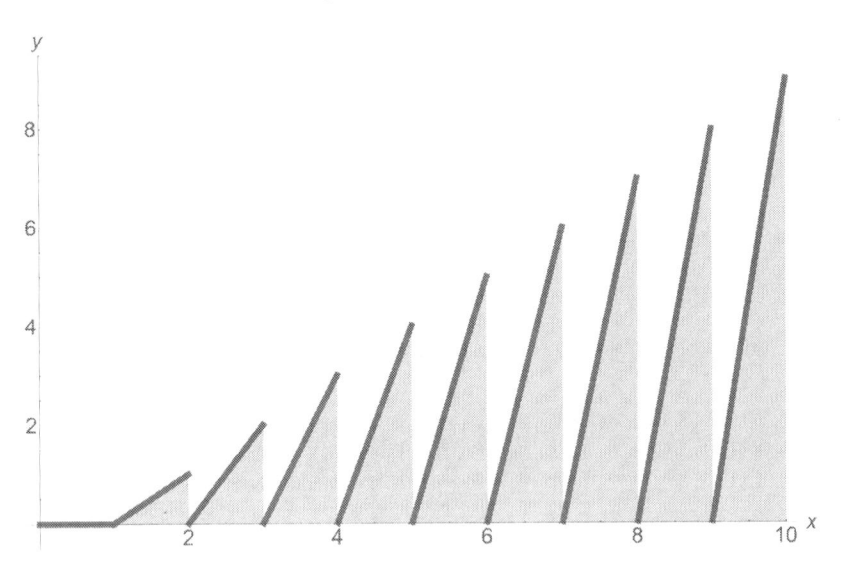

Figure 8.2: Try to decipher the pattern from this graph to solve the integral.

Recall the definition of the fractional part:

$$\{x\} = x - \lfloor x \rfloor$$

8.2. SOME PROBLEMS

From the previous example, we know that for some $x \in [k, k+1)$, $\{x\} = x - k$. We also know that on that same interval, $\lfloor x \rfloor = k$. Therefore,

$$I = \int_0^n \{x\}\lfloor x \rfloor \mathrm{d}x = \int_0^1 (x-0)(0) \, \mathrm{d}x + \int_1^2 (x-1)(1) \, \mathrm{d}x + \cdots$$

$$= \sum_{k=0}^{n-1} \int_k^{k+1} k(x-k) \, \mathrm{d}x$$

$$= \sum_{k=0}^{n-1} \left[\frac{kx^2}{2} - k^2 x \right]_k^{k+1}$$

$$= \sum_{k=0}^{n-1} \frac{k}{2}$$

$$= \frac{1}{2} \sum_{k=0}^{n-1} k = \frac{n(n-1)}{4}$$

Example 3: Evaluate $\displaystyle\int_0^\infty \frac{\{x\}^{\lfloor x \rfloor}}{\lceil x \rceil} \mathrm{d}x$

Solution

As always, we will break this integral into a sum of integrals. However, now we are dealing with an infinite sum!

$$I = \sum_{n=0}^{\infty} \int_n^{n+1} \frac{\{x\}^{\lfloor x \rfloor}}{\lceil x \rceil} \mathrm{d}x$$

$$= \sum_{n=0}^{\infty} \int_n^{n+1} \frac{\{x\}^n}{n+1} \mathrm{d}x$$

238   CHAPTER 8. FRACTIONAL PART INTEGRALS

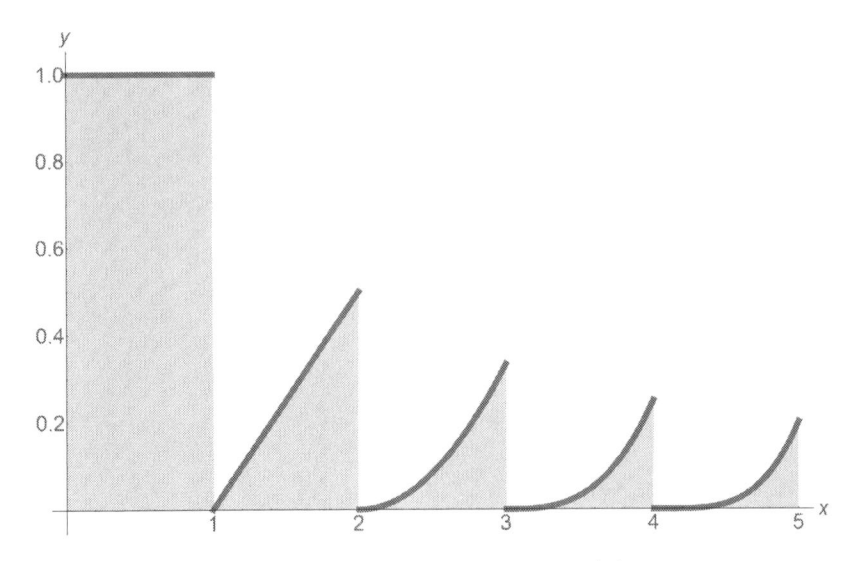

Figure 8.3: Graph of $y = \frac{\{x\}^{\lfloor x \rfloor}}{\lceil x \rceil}$

$$= \sum_{n=0}^{\infty} \int_n^{n+1} \frac{(x-n)^n}{n+1} \, dx$$

Notice that we can substitute $x - n \to x$ to get:

$$I = \sum_{n=0}^{\infty} \int_0^1 \frac{x^n}{n+1} dx$$

$$= \sum_{n=0}^{\infty} \frac{1}{(n+1)^2}$$

$$\therefore I = \sum_{n=1}^{\infty} \frac{1}{n^2} = \zeta(2) = \frac{\pi^2}{6}$$

Example 4: Evaluate $\int_0^1 \left\{\frac{1}{x}\right\} dx$

8.2. SOME PROBLEMS

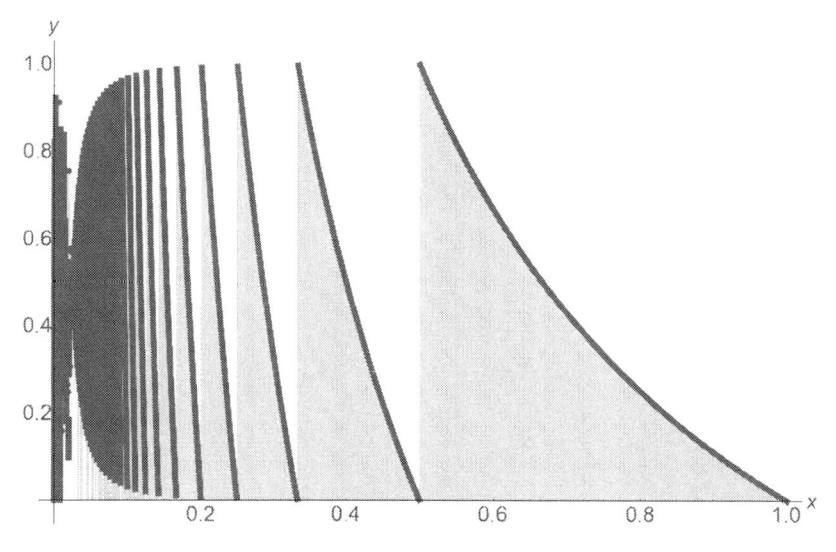

Figure 8.4: Graph of $y = \left\{\frac{1}{x}\right\}$

This integral is part of a well-known and beautiful result in mathematics[1]. We begin solving this integral by substituting $u = \frac{1}{x}$, $dx = -\frac{du}{u^2}$. We then have

$$I = \int_0^1 \left\{\frac{1}{x}\right\} dx$$

$$= \int_1^\infty \frac{\{u\}}{u^2} du$$

Notice that we can break down the integral above into a consecutive sum of integrals.

$$I = \sum_{n=1}^\infty \int_n^{n+1} \frac{u-n}{u^2} du$$

---

[1] Havil, J. *Gamma: Exploring Euler's Constant.* Princeton, NJ: Princeton University Press, pp. 109-111, 2003.

$$= \sum_{n=1}^{\infty} \left(\int_n^{n+1} \frac{du}{u} - n \int_n^{n+1} \frac{du}{u^2} \right)$$

Which evaluates to

$$I = \sum_{k=1}^{\infty} \left(\ln\left(\frac{n+1}{n}\right) - \frac{1}{n+1} \right)$$

Recall the definition of the Euler-Mascheroni constant (See (1.8)):

$$\gamma = \lim_{k \to \infty} (H_k - \ln k)$$

We can replace $\ln k$ by $\ln(k+1)$ since $\lim_{k \to \infty} \ln(k+1) - \ln k = 0$. Thus,

$$\gamma = \lim_{k \to \infty} (H_k - \ln(k+1))$$

Notice that $\ln(k+1)$ can be represented by a telescoping sum:

$$\ln(k+1) = \sum_{n=1}^{k} \ln(n+1) - \ln n = \sum_{n=1}^{k} \ln\left(\frac{n+1}{n}\right)$$

Plugging the sum above into the expression for γ then gives

$$\gamma = \lim_{k \to \infty} \left[\sum_{n=1}^{k} \frac{1}{n} - \sum_{n=1}^{k} \ln\left(\frac{n+1}{n}\right) \right]$$

$$= \lim_{k \to \infty} \sum_{n=1}^{k} \left(\frac{1}{n} - \ln\left(\frac{n+1}{n}\right) \right)$$

8.2. SOME PROBLEMS

This well-known result is due to Euler. The expression we obtained for I is then:

$$I = \sum_{k=1}^{\infty}\left(\ln\left(\frac{n+1}{n}\right) - \frac{1}{n+1}\right) = 1 - \gamma \qquad (8.4)$$

Example 5: Evaluate $\int_0^1 \left\{\frac{1}{\sqrt[3]{x}}\right\} dx$

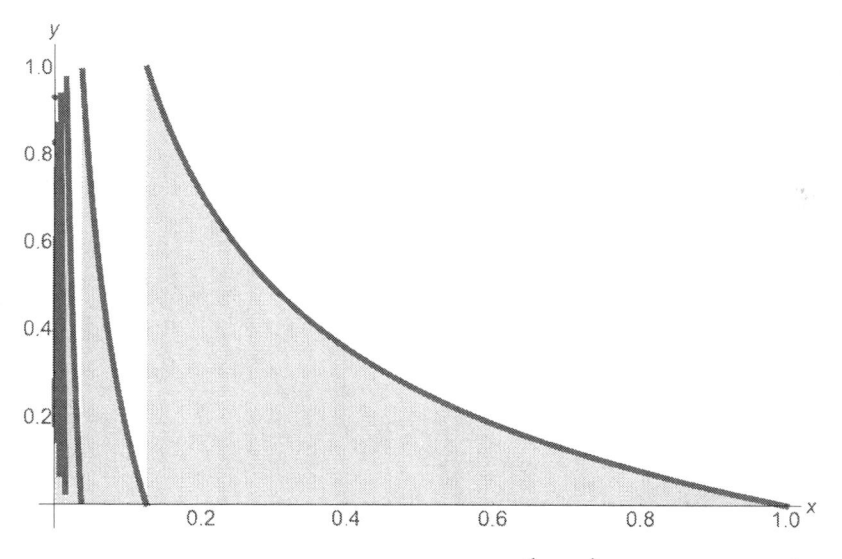

Figure 8.5: Graph $y = \left\{\frac{1}{\sqrt[3]{x}}\right\}$

We can generalize this for any root. Define

$$I_k = \int_0^1 \left\{\frac{1}{\sqrt[k]{x}}\right\} dx$$

Where $k > 1$. Substituting $x = \frac{1}{t^k}$, $dx = -\frac{k}{t^{k+1}} du$ gives

CHAPTER 8. FRACTIONAL PART INTEGRALS

$$I_k = k \int_1^\infty \frac{\{t\}}{t^{k+1}} dt$$

As usual, we will break this integral up into a sum of consecutive integrals

$$I_k = k \sum_{n=1}^\infty \int_n^{n+1} \frac{t-n}{t^{k+1}} dt$$

Evaluating the integral and simplifying a bit,

$$I_k = k \left[\frac{1}{1-k} \sum_{n=1}^\infty \left(\frac{1}{(n+1)^{k-1}} - \frac{1}{n^{k-1}} \right) \right.$$

$$\left. + \frac{1}{k} \sum_{n=1}^\infty n \left(\frac{1}{(n+1)^k} - \frac{1}{n^k} \right) \right]$$

Notice that the first sum is telescoping:

$$\sum_{n=1}^\infty \left(\frac{1}{(n+1)^{k-1}} - \frac{1}{n^{k-1}} \right) = \left(\frac{1}{2^{k-1}} - \frac{1}{1^{k-1}} \right) +$$

$$\left(\frac{1}{3^{k-1}} - \frac{1}{2^{k-1}} \right) + \left(\frac{1}{4^{k-1}} - \frac{1}{3^{k-1}} \right) + \cdots$$

$$= \lim_{n \to \infty} \left(-\frac{1}{1^{k-1}} + \frac{1}{n^{k-1}} \right)$$

$$= -1$$

Therefore,

8.2. SOME PROBLEMS

$$I_k = \frac{k}{k-1} + \sum_{n=1}^{\infty} n \left(\frac{1}{(n+1)^k} - \frac{1}{n^k} \right)$$

Notice that the sum above is

$$S = \sum_{n=1}^{\infty} n \left(\frac{1}{(n+1)^k} - \frac{1}{n^k} \right) = 1 \left(\frac{1}{2^k} - \frac{1}{1^k} \right) + 2 \left(\frac{1}{3^k} - \frac{1}{2^k} \right)$$
$$+ 3 \left(\frac{1}{4^k} - \frac{1}{3^k} \right) + \cdots$$

We can regroup the terms in the sum without changing its value (Since it is absolutely convergent for $k > 1$):

$$S = \left(-\frac{1}{1^k} \right) + \left(\frac{1}{2^k} - 2 \cdot \frac{1}{2^k} \right) + \left(2 \cdot \frac{1}{3^k} - 3 \cdot \frac{1}{3^k} \right) + \cdots$$

Each $\frac{n-1}{n^k}$ gets $\frac{n}{n^k}$ subtracted from it. We therefore have:

$$S = - \left(\frac{1}{1^k} + \frac{1}{2^k} + \frac{1}{3^k} + \cdots \right)$$
$$= -\zeta(k)$$

Plugging S back in we get

$$\int_0^1 \left\{ \frac{1}{\sqrt[k]{x}} \right\} dx = \frac{k}{k-1} - \zeta(k) \qquad (8.5)$$

To evaluate our integral, we simply substitute $k = 3$ to obtain:

$$\int_0^1 \left\{ \frac{1}{\sqrt[3]{x}} \right\} dx = \frac{3}{2} - \zeta(3)$$

244   CHAPTER 8. FRACTIONAL PART INTEGRALS

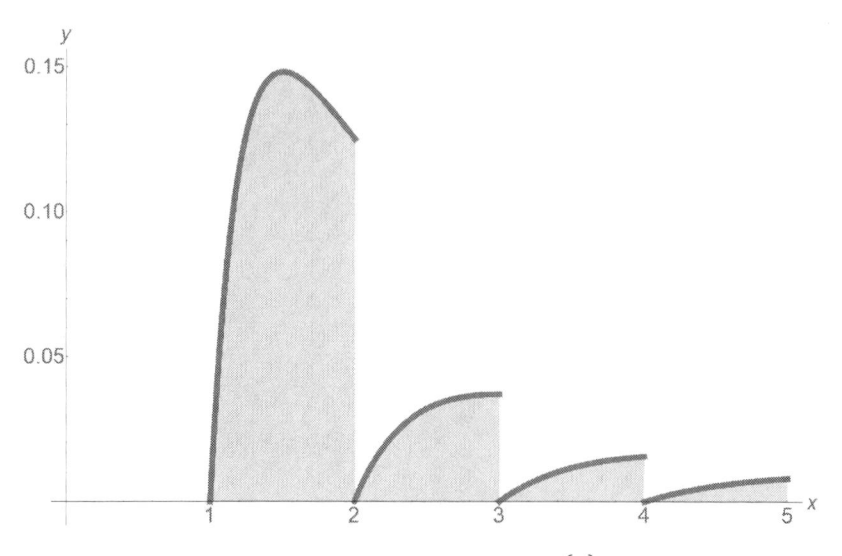

Figure 8.6: Graph of $y = \frac{\{x\}}{x^3}$

Example 6: Evaluate $\displaystyle\int_1^\infty \frac{\{x\}}{x^n}\mathrm{d}x$ for $n > 2$.

We begin by substituting $\{x\} = x - \lfloor x \rfloor$.

$$\implies I = \int_1^\infty \frac{x - \lfloor x \rfloor}{x^n}\mathrm{d}x$$

$$= \int_1^\infty \frac{\mathrm{d}x}{x^{n-1}} - \int_1^\infty \frac{\lfloor x \rfloor}{x^n}\mathrm{d}x$$

The first integral is trivial to evaluate. Notice that we can rewrite the second integral as an infinite sum of consecutive integrals

$$I = \frac{1}{n-2} - \sum_{k=1}^\infty \int_k^{k+1} \frac{k}{x^n}\mathrm{d}x$$

We have already evaluated a similar integral in (8.5). Therefore,

8.2. SOME PROBLEMS

$$I = \int_1^\infty \frac{\{x\}}{x^n} dx = \frac{1}{n-2} - \frac{\zeta(n-1)}{n-1} \tag{8.6}$$

Example 7: Evaluate $\int_0^1 \left\{\frac{1}{k\sqrt[k]{x}}\right\} dx$ for $k > 1$.

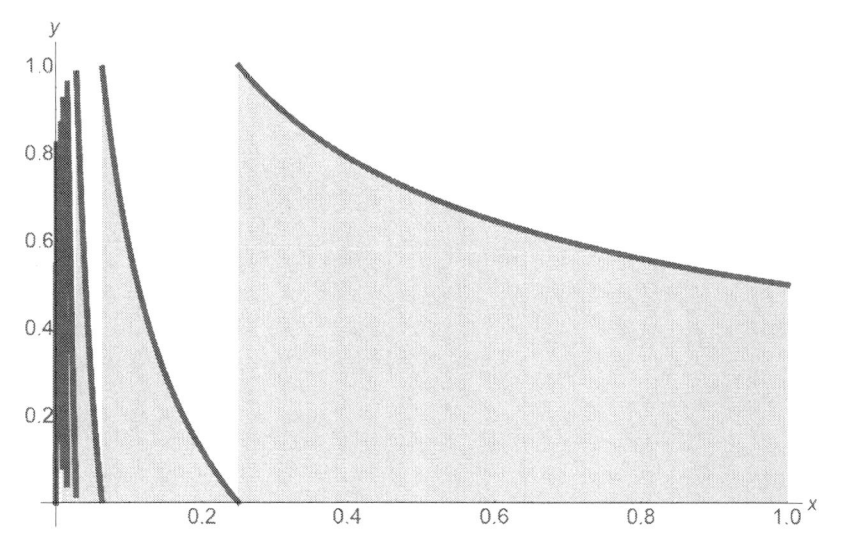

Figure 8.7: Graph of $y = \left\{\frac{1}{2\sqrt{x}}\right\}$

We begin by substituting $u = k\sqrt[k]{x}$, $dx = \frac{u^{k-1}}{k^{k-1}} du$,

$$I = \int_0^1 \left\{\frac{1}{k\sqrt[k]{x}}\right\} dx = \frac{1}{k^{k-1}} \int_0^k \left\{\frac{1}{u}\right\} u^{k-1} du$$

Substituting $y = \frac{1}{u}$, $du = -\frac{dy}{y^2}$,

$$I = \frac{1}{k^{k-1}} \int_{\frac{1}{k}}^\infty \frac{\{y\}}{y^{k+1}} dy$$

Recall the integral we dealt with in (8.5),

$$I_k = \int_0^1 \left\{ \frac{1}{\sqrt[k]{x}} \right\} dx$$

Which transforms into

$$I_k = k \int_1^\infty \frac{\{t\}}{t^{k+1}} dt = \frac{k}{k-1} - \zeta(k)$$

Through the substitution $x = \frac{1}{t^k}$. We can then use our result from that example to obtain:

$$I = \frac{1}{k^{k-1}} \int_{\frac{1}{k}}^\infty \frac{\{y\}}{y^{k+1}} dy$$

$$= \frac{1}{k^{k-1}} \left[\int_{\frac{1}{k}}^1 \frac{\{y\}}{y^{k+1}} dy + \underbrace{\int_1^\infty \frac{\{y\}}{y^{k+1}} dy}_{I_k/k} \right]$$

$$= \frac{1}{k^{k-1}} \left[\int_{\frac{1}{k}}^1 \frac{\{y\}}{y^{k+1}} dy + \frac{1}{k-1} - \frac{\zeta(k)}{k} \right]$$

Notice that for $y \in \left[\frac{1}{k}, 1\right)$, $\{y\} = y$. Therefore,

$$I = \frac{1}{k^{k-1}} \left[\int_{\frac{1}{k}}^1 \frac{dy}{y^k} + \frac{1}{k-1} - \frac{\zeta(k)}{k} \right]$$

$$= \frac{1}{k^{k-1}} \left[\frac{k^{k-1} - 1}{k-1} + \frac{1}{k-1} - \frac{\zeta(k)}{k} \right]$$

$$= \frac{1}{k^{k-1}} \left[\frac{k^{k-1}}{k-1} - \frac{\zeta(k)}{k} \right]$$

8.2. SOME PROBLEMS

Which finally evaluates to

$$I = \int_0^1 \left\{\frac{1}{k\sqrt[k]{x}}\right\} dx = \frac{1}{k-1} - \frac{\zeta(k)}{k^k}$$

Example 8: Evaluate $\int_0^1 \frac{\{2x\}}{x} \left\{\frac{1}{x}\right\} dx$

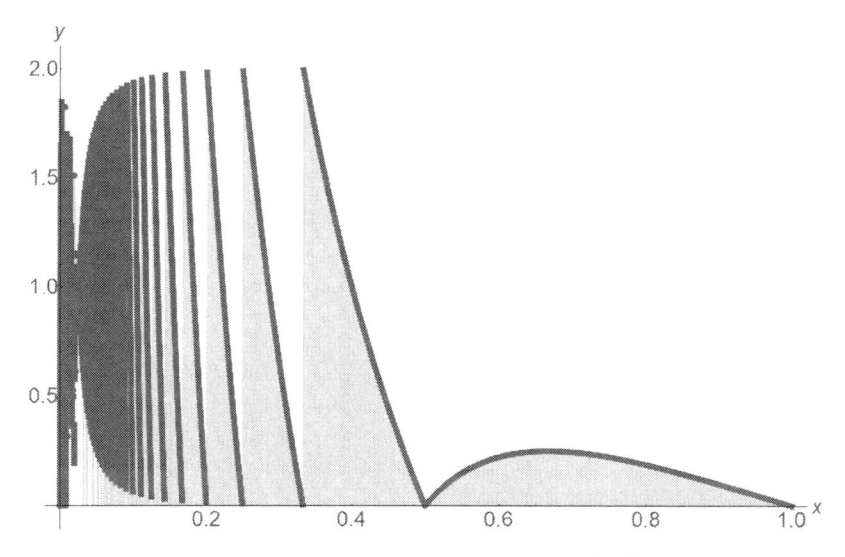

Figure 8.8: Graph of $y = \frac{\{2x\}}{x} \left\{\frac{1}{x}\right\}$

Notice that when $x \in \left[0, \frac{1}{2}\right)$, $\{2x\} = 2x$ and when $x \in \left[\frac{1}{2}, 1\right)$, $\{2x\} = 2x - 1$. Thus, we can split our desired integral into two separate integrals:

$$I = \int_0^1 \left\{\frac{1}{x}\right\} \{2x\} \frac{dx}{x}$$

$$= \int_0^{\frac{1}{2}} (2x) \left\{\frac{1}{x}\right\} \frac{dx}{x} + \int_{\frac{1}{2}}^1 (2x - 1) \left\{\frac{1}{x}\right\} \frac{dx}{x}$$

$$= 2\int_0^{\frac{1}{2}} \left\{\frac{1}{x}\right\} dx + 2\int_{\frac{1}{2}}^1 \left\{\frac{1}{x}\right\} dx - \int_{\frac{1}{2}}^1 \left\{\frac{1}{x}\right\} \frac{dx}{x}$$

We can combine the first two integrals to get:

$$I = 2\int_0^1 \left\{\frac{1}{x}\right\} dx - \int_{\frac{1}{2}}^1 \left\{\frac{1}{x}\right\} \frac{dx}{x}$$

We have already evaluated the first integral in (8.4). In regards to the second integral, notice that when $x \in \left(\frac{1}{2}, 1\right]$, $\left\{\frac{1}{x}\right\} = \frac{1}{x} - 1$ by the definition of the fractional part function. Thus,

$$\int_{\frac{1}{2}}^1 \left\{\frac{1}{x}\right\} \frac{dx}{x} = \int_{\frac{1}{2}}^1 \frac{\left(\frac{1}{x} - 1\right)}{x} dx$$

$$= \int_{\frac{1}{2}}^1 \frac{dx}{x^2} - \int_{\frac{1}{2}}^1 \frac{dx}{x}$$

$$= 1 - \ln 2$$

Thus,

$$I = 2(1 - \gamma) - (1 - \ln 2)$$

$$= 1 + \ln 2 - 2\gamma$$

Example 9: Evaluate $\int_1^\infty \frac{\{x\} - \frac{1}{2}}{x} dx$

8.2. SOME PROBLEMS

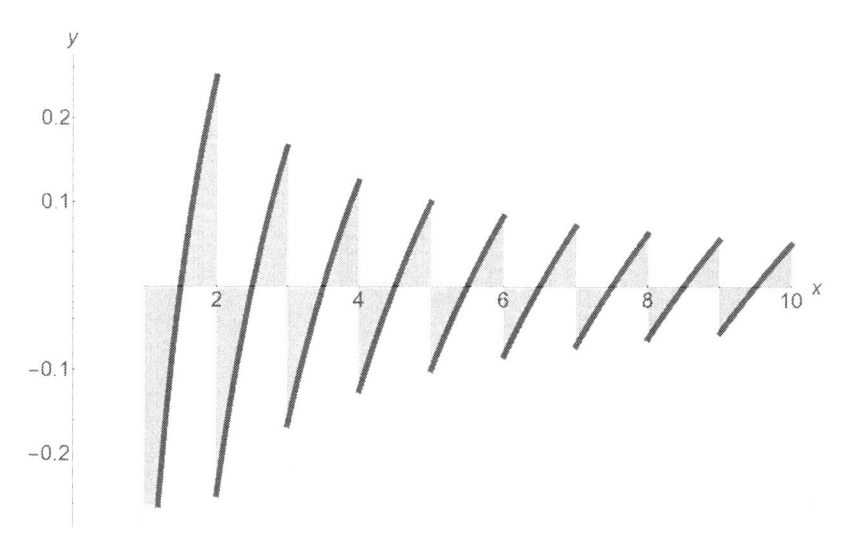

Figure 8.9: Graph of $y = \frac{\{x\}-\frac{1}{2}}{x}$

By the definition of the fractional part function,

$$I = \int_1^\infty \frac{x - \lfloor x \rfloor - \frac{1}{2}}{x} \, \mathrm{d}x$$

We can transform this integral into a sum of integrals with intervals $[k, k+1]$ so we can get rid of the floor function

$$I = \sum_{k=1}^\infty \int_k^{k+1} \frac{x - \lfloor x \rfloor - \frac{1}{2}}{x} \, \mathrm{d}x$$

$$= \sum_{k=1}^\infty \int_k^{k+1} \frac{x - k - \frac{1}{2}}{x} \, \mathrm{d}x$$

In evaluating the integral above we obtain

$$I = \sum_{k=1}^{\infty} \left[\left(k + \frac{1}{2}\right) \ln k - \left(k + \frac{1}{2}\right) \ln(k+1) + 1 \right]$$

Notice that we can add $\ln k - \ln k$ to get:

$$I = \sum_{k=1}^{\infty} \left[\left(k + 1 - 1 + \frac{1}{2}\right) \ln k - \left(k + \frac{1}{2}\right) \ln(k+1) + 1 \right]$$

$$= \lim_{N \to \infty} \sum_{k=1}^{N} \left[1 + \ln k + \left(k - 1 + \frac{1}{2}\right) \ln k - \left(k + \frac{1}{2}\right) \ln(k+1) \right]$$
(8.7)

We can split (8.7) into two the limit of two summations:

$$I = \lim_{N \to \infty} \left[\sum_{k=1}^{N} (1 + \ln k) \right.$$

$$\left. + \underbrace{\sum_{k=1}^{N} \left(\left(k - 1 + \frac{1}{2}\right) \ln k - \left(k + \frac{1}{2}\right) \ln(k+1) \right)}_{\text{Telescoping Sum}} \right]$$

$$= \lim_{N \to \infty} \left(N + \ln N! - \left(N + \frac{1}{2}\right) \ln(N+1) \right)$$

This looks like an opportunity to use Stirling's formula (See (1.6)). However, we first need to convert our expression into one logarithm.

$$I = \lim_{N \to \infty} \left(\ln\left(e^N\right) + \ln N! - \ln\left((N+1)^{N+\frac{1}{2}}\right) \right)$$

8.2. SOME PROBLEMS

$$= \lim_{N\to\infty} \ln\left(\frac{N!\, e^N}{(N+1)^{N+\frac{1}{2}}}\right)$$

Since $N \to \infty$, we can use (1.6) to get

$$I = \lim_{N\to\infty} \ln\left(\frac{e^N \left(\frac{N}{e}\right)^N \sqrt{2\pi N}}{(N+1)^{N+\frac{1}{2}}}\right)$$

$$= \lim_{N\to\infty} \ln\left(\frac{N^{N+\frac{1}{2}}\sqrt{2\pi}}{(N+1)^{N+\frac{1}{2}}}\right)$$

$$= \lim_{N\to\infty} \ln\left(\frac{\sqrt{2\pi}}{\left(1+\frac{1}{N}\right)^{N+\frac{1}{2}}}\right)$$

By the definition of e,

$$e = \lim_{x\to\infty} \left(1+\frac{1}{x}\right)^x$$

We have

$$I = \lim_{N\to\infty} \ln\left(\frac{\sqrt{2\pi}}{e}\right)$$

$$= \ln\left(\frac{\sqrt{2\pi}}{e}\right)$$

We finally obtain that:

$$\int_1^\infty \frac{\{x\}-\frac{1}{2}}{x}\,\mathrm{d}x = \ln\left(\sqrt{2\pi}\right) - 1$$

CHAPTER 8. FRACTIONAL PART INTEGRALS

Example 10: Evaluate $\displaystyle\int_0^1 \left\{\frac{1}{x}\right\} x \ln x \, dx$

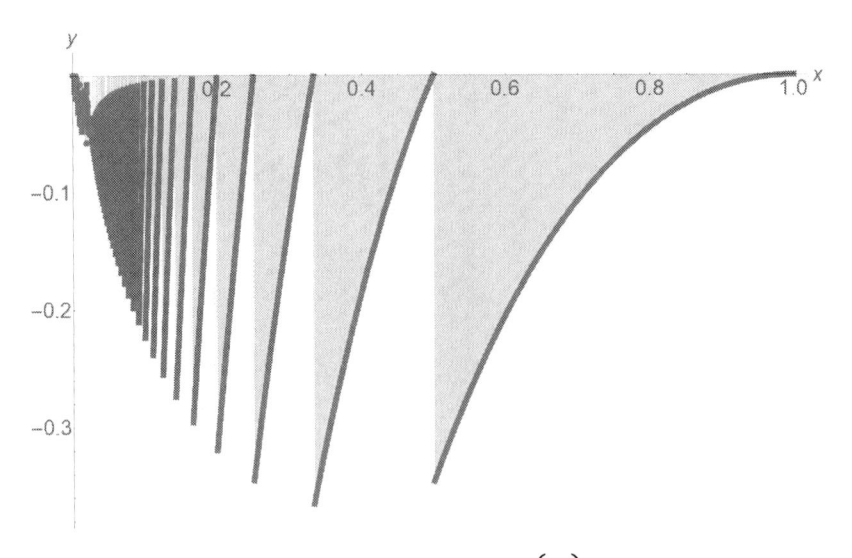

Figure 8.10: Graph of $y = \left\{\frac{1}{x}\right\} x \ln x$

Solution

Define a function

$$f(\alpha) = \int_0^1 \left\{\frac{1}{x}\right\} x^{\alpha-1} dx$$

Differentiating under the integral,

$$f'(\alpha) = \int_0^1 \left\{\frac{1}{x}\right\} x^{\alpha-1} \ln x \, dx$$

Our desired integral is then $f'(2)$. Let us first evaluate $f(\alpha)$. Consider the substitution $u = \frac{1}{x}$,

8.2. SOME PROBLEMS

$$\implies f(\alpha) = \int_0^\infty \frac{\{x\}}{x^{\alpha+1}}\,dx$$

We have already evaluated this integral (See (8.6)). Therefore,

$$f(\alpha) = \frac{1}{\alpha - 1} - \frac{\zeta(\alpha)}{\alpha}$$

And,

$$f'(\alpha) = -\frac{1}{(\alpha-1)^2} - \frac{\alpha\zeta'(\alpha) - \zeta(\alpha)}{\alpha^2}$$

$$= \frac{\zeta(\alpha)}{\alpha^2} - \frac{1}{(\alpha-1)^2} - \frac{\zeta'(\alpha)}{\alpha}$$

Plugging in $\alpha = 2$ gives

$$f(2) = \frac{\zeta(2)}{4} - 1 - \frac{\zeta'(2)}{2}$$

The value of $\zeta'(2)$ is:

$$\zeta'(2) = -\sum_{k=1}^\infty \frac{\ln k}{k^2} = -\frac{\pi^2}{6}\left(12\ln A - \gamma - \ln(2\pi)\right) \qquad (8.8)$$

Where A denotes the Glaisher-Kinkelin constant.

> **Definition**
>
> The Glaisher-Kinkelin constant, usually denoted as A, is a constant that appears in many sums and integrals. It is be given by:
>
> $$A = \lim_{n \to \infty} \frac{K(n+1)}{e^{-n^2/4} n^{n^2/2+n/2+1/12}} \approx 1.282 \qquad (8.9)$$
>
> Where $K(\cdot)$ denotes the K-function:
>
> $$K(n) = \prod_{k=1}^{n-1} k^k$$

(8.8) can be easily shown through Glaisher's 1894 result:[2]

$$\prod_{k=1}^{\infty} k^{\frac{1}{k^2}} = \left(\frac{A^{12}}{2\pi e^{\gamma}}\right)^{\frac{\pi^2}{6}} \qquad (8.10)$$

Thus,

$$f(2) = \frac{\pi^2}{24} - 1 - \frac{\pi^2}{12}\left(12\ln A - \gamma - \ln(2\pi)\right)$$

---

[2] Glaisher, J. W. L. *On the Constant which Occurs in the Formula for $1^1.2^2.3^3 \ldots n^n$*. Messenger Math. 24, 1-16, 1894.

8.3 Open Problems

> **Open Problem**
>
> Let $n \geq 3$ be an integer. Calculate
> $$\int_{[0,1]^n} \left\{\frac{x_1}{x_2}\right\} \left\{\frac{x_2}{x_3}\right\} \cdots \left\{\frac{x_n}{x_1}\right\} \mathrm{d}x_1 \cdots \mathrm{d}x_n$$
> Where
> $$\int_{[0,1]^n} \equiv \underbrace{\int_0^1 \cdots \int_0^1}_{n \text{ times}}$$

> **Open Problem**
>
> Let $n \geq 3, k \geq 1$ be integers. Calculate
> $$\int_{[0,1]^n} \left\{\frac{1}{x_1 + x_2 + \cdots + x_n}\right\}^k \mathrm{d}x_1 \cdots \mathrm{d}x_n$$

Where the open problems are due to Ovidiu Furdui's book "Limits, Series, and Fractional Part Integrals: Problems in Mathematical Analysis"[3]. It is worth mentioning that many of the fractional part integrals outlined in this chapter can be found in Furdui's book, along with solutions.

8.4 Exercise Problems

1) Evaluate $\int_0^n \{x^2\} \, \mathrm{d}x$ for $n \in \mathbb{N}$.

---

[3] Furdui, O. (2013). *Limits, Series, and Fractional Part Integrals: Problems in Mathematical Analysis*. New York, NY: Springer.

2) Calculate $\displaystyle\int_0^1 \int_0^1 \left\{\frac{x+y}{x-y}\right\} dx\, dy$

3) Find a closed form for $\displaystyle\int_0^1 \int_0^1 (xy)^{2019} \left\{\frac{x}{y}\right\} \left\{\frac{y}{x}\right\} dx\, dy$

4) Evaluate $\displaystyle\int_0^1 x \left\lfloor \frac{1}{x} \right\rfloor \left\{\frac{1}{x}\right\} dx$

5)

> **Challenge Problem**
>
> Evaluate
>
> $$\int_0^1 \sqrt{\frac{\left\{\frac{1}{x}\right\}}{1 - \left\{\frac{1}{x}\right\}^{1-x}}}\, dx$$
>
> Hint: Use the integral definition of the gamma function.

Part III

A Study in the Special Functions

In this part, we will use our tools from the previous chapters to derive properties and representations of many special functions. After doing so, we will work through a collection of example problems with the aid of our new powerful tools.

This part will involve extensive use of the gamma, polygamma, beta, and Riemann zeta functions. Our aim is to explore their utility in series and integral problems as well as to establish a sense of familiarity with their properties and uses. These tools will then be used in our final and culminating part of the book regarding applications in the mathematical sciences.

Chapter 9

Gamma Function

9.1 Definition

We have already introduced this function back in chapter 1 (See (1.7)). This special function is a very powerful tool in evaluating many integrals and series, and is perhaps one of the most applicable special functions in both mathematics and the mathematical sciences.

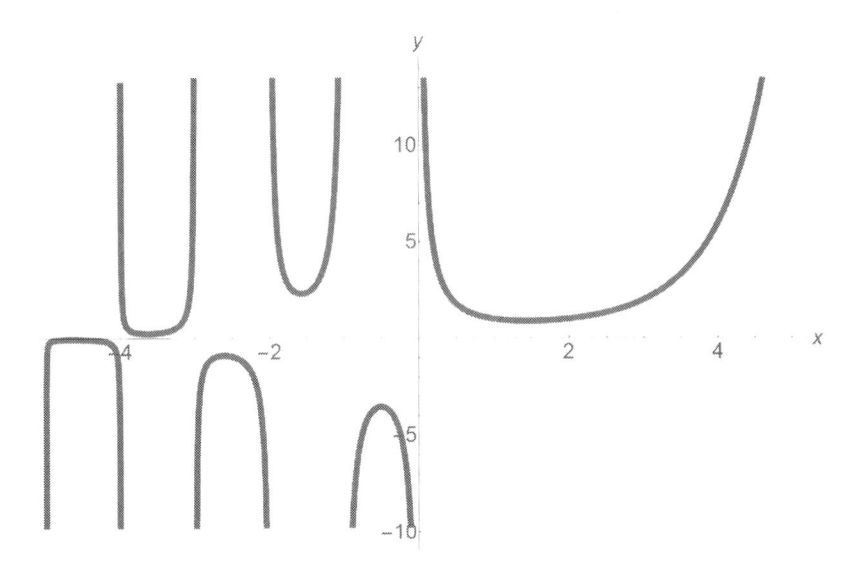

Figure 9.1: Graph of $y = \Gamma(x)$

9.2 Special Values

9.3. PROPERTIES AND REPRESENTATIONS

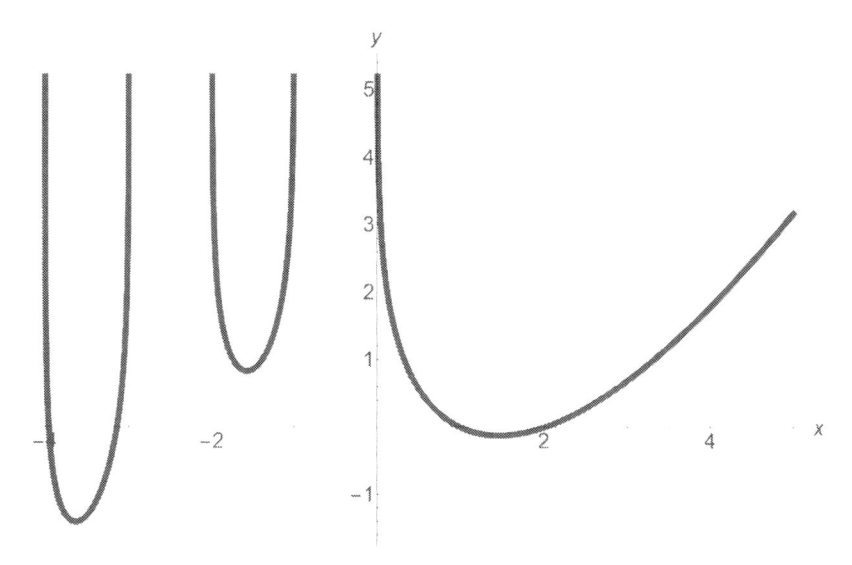

Figure 9.2: Graph of the log-gamma function, $y = \ln \Gamma(x)$, which arises in various problems in mathematical analysis

$$\Gamma(1) = 0! = 1$$

$$\Gamma\left(\frac{1}{2}\right) = \sqrt{\pi}$$

$$\Gamma\left(-\frac{1}{2}\right) = -2\sqrt{\pi}$$

$$\Gamma\left(\frac{1}{3}\right) \approx 2.678\,938\,534\,707\,747\,6337$$

$$\Gamma\left(\frac{1}{4}\right) \approx 3.625\,609\,908\,221\,908\,3119$$

9.3 Properties and Representations

In the properties and representations section, we will derive several identities and equations relating to the special function considered in the chapter. Let us get started with our first!

> **Theorem**
>
> An interesting equation due to Gauss:
>
> $$\Gamma(z) = \lim_{n\to\infty} \frac{n^z}{z} \prod_{k=1}^{n} \frac{k}{z+k} \qquad (9.1)$$

Proof. Denote
$$I_n = \int_0^n t^{z-1}\left(1 - \frac{t}{n}\right)^n dt$$

Applying IBP with $u = \left(1 - \frac{t}{n}\right)^n$, $dv = t^{z-1} dt$ gives

$$I_n = \underbrace{\left[\frac{t^z}{z}\left(1 - \frac{t}{n}\right)^n\right]_0^n}_{=0} + \frac{n}{nz} \int_0^n t^z \left(1 - \frac{t}{n}\right)^{n-1} dt$$

$$= \frac{n}{nz} \int_0^n t^z \left(1 - \frac{t}{n}\right)^{n-1} dt$$

Applying IBP successively with $u = \left(1 - \frac{t}{n}\right)^{n-k}$, $dv = t^{z+k-1} dt$ for $k = 1, 2, 3 \cdots, n-1$ gives the formula

$$I_n = \frac{n}{nz} \cdot \frac{n-1}{n(z+1)} \cdots \frac{1}{n(z+n-1)} \int_0^n t^{z+n-1} dt$$

$$= \frac{n}{nz} \cdot \frac{n-1}{n(z+1)} \cdots \frac{1}{n(z+n-1)} \cdot \frac{n^{n+z}}{n+z}$$

Note that there are n multiples of n in the denominator. We can easily express the above expression as a finite product:

9.3. PROPERTIES AND REPRESENTATIONS

$$I_n = \frac{n^z}{z} \prod_{k=1}^{n} \frac{k}{z+k}$$

Since

$$\lim_{n \to \infty} I_n = \lim_{n \to \infty} \int_0^n t^{z-1} \left(1 - \frac{t}{n}\right)^n dt$$

And

$$e^{-t} = \lim_{n \to \infty} \left(1 - \frac{t}{n}\right)^n$$

We can write:

$$\lim_{n \to \infty} I_n = \int_0^\infty t^{z-1} e^{-t} dt = \Gamma(z)$$

Where the interchange of the limit and integral is justified by the dominated convergence theorem. Thus,

$$\Gamma(z) = \lim_{n \to \infty} \frac{n^z}{z} \prod_{k=1}^{n} \frac{k}{z+k}$$

□

> **Theorem**
>
> Another infinite product expression for the gamma function is given by Weierstrass:
>
> $$\Gamma(z) = \frac{e^{-\gamma z}}{z} \prod_{k=1}^{\infty} \left(1 + \frac{z}{k}\right)^{-1} e^{z/k} \qquad (9.2)$$

Proof. Consider the result from (9.1):

$$\Gamma(z) = \lim_{n \to \infty} \frac{n^z}{z} \prod_{k=1}^{n} \frac{k}{z+k}$$

We can rewrite this as

$$\Gamma(z) = \lim_{n \to \infty} \frac{1}{z} \exp\left(zH_n - zH_n + z\ln(n)\right) \prod_{k=1}^{n} \frac{k}{z+k}$$

Since

$$\lim_{n \to \infty} -\ln n + H_n = \gamma$$

We can express our product as

$$\Gamma(z) = \lim_{n \to \infty} \frac{1}{z} \exp\left(zH_n - z\gamma\right) \prod_{k=1}^{n} \frac{k}{z+k}$$

$$= \lim_{n \to \infty} \frac{1}{z} \exp\left(zH_n - z\gamma\right) \prod_{k=1}^{n} \frac{1}{1+\frac{z}{k}}$$

$$= \lim_{n \to \infty} \frac{e^{-\gamma z}}{z} \exp\left(zH_n\right) \prod_{k=1}^{n} \frac{1}{1+\frac{z}{k}}$$

Notice that

$$\exp\left(zH_n\right) = \exp\left(z\sum_{k=1}^{n} \frac{1}{k}\right)$$

$$= e^{\frac{z}{1}} \cdot e^{\frac{z}{2}} \cdots e^{\frac{z}{n}}$$

9.3. PROPERTIES AND REPRESENTATIONS

$$= \prod_{k=1}^{n} e^{\frac{z}{k}}$$

Thus,

$$\Gamma(z) = \lim_{n\to\infty} \frac{e^{-\gamma z}}{z} \prod_{k=1}^{n} e^{\frac{z}{k}} \prod_{k=1}^{n} \frac{1}{1+\frac{z}{k}}$$

$$= \lim_{n\to\infty} \frac{e^{-\gamma z}}{z} \prod_{k=1}^{n} \frac{e^{\frac{z}{k}}}{1+\frac{z}{k}}$$

Taking the limit as $n \to \infty$,

$$\Gamma(z) = \frac{e^{-\gamma z}}{z} \prod_{k=1}^{\infty} \left(1+\frac{z}{k}\right)^{-1} e^{z/k}$$

□

> **Theorem**
>
> The infamous Euler's reflection formula is given by
>
> $$\Gamma(z)\Gamma(1-z) = \pi \csc \pi z \qquad (9.3)$$
>
> This equation has been used multiple times throughout the book, so why not prove it!

Proof. Consider equation (9.1):

$$\Gamma(z) = \lim_{n\to\infty} \frac{n^z}{z} \prod_{k=1}^{n} \frac{k}{z+k} \qquad (9.4)$$

The substitution $z \to -z$ gives

$$\Gamma(-z) = \lim_{n \to \infty} \frac{1}{z \cdot n^z} \prod_{k=1}^{n} \frac{k}{z-k} \qquad (9.5)$$

We can then multiply (9.4) and (9.5) to get

$$\Gamma(z)\Gamma(-z) = \lim_{n \to \infty} \frac{1}{z^2} \prod_{k=1}^{n} \frac{k^2}{z^2 - k^2}$$

$$= \lim_{n \to \infty} \frac{1}{z^2} \prod_{k=1}^{n} \frac{1}{\frac{z^2}{k^2} - 1} \qquad (9.6)$$

Lemma. We can express $\sin \pi x$ as

$$\sin \pi x = \pi x \prod_{k=1}^{\infty} \left(1 - \frac{x^2}{k^2}\right) \qquad (9.7)$$

Proof. (9.7) was used by Euler to prove the famous result that

$$\zeta(2) = \sum_{n=1}^{\infty} \frac{1}{n^2} = \frac{\pi^2}{6}$$

The problem of evaluating the above sum is widely known as the *Basel problem*. The problem was originally posed by Pietro Mengoli in 1650, and gained notoriety after many famous mathematicians failed to attack it, notably the Bernoulli family[1]. In this proof, we will present the heuristic approach Euler took, which was only justified 100 years later by Weierstrass.

In his infamous solution of the Basel problem, Euler argued heuristically that one can express $\frac{\sin x}{x}$ as a polynomial of infinite degree based on its roots, similar to how one can factorize a finite polynomial. In doing so, Euler gave

---

[1] Ayoub, Raymond (1974). *Euler and the zeta function*. Amer. Math. Monthly. 81: 1067–86. doi:10.2307/2319041.

9.3. PROPERTIES AND REPRESENTATIONS

$$\frac{\sin x}{x} = (x+\pi)(x-\pi)(x+2\pi)(x-2\pi)\cdots$$

$$= (x^2 - \pi^2)(x^2 - 4\pi^2)(x^2 - 9\pi^2)\cdots$$

$$= A\left(1 - \frac{x^2}{\pi^2}\right)\left(1 - \frac{x^2}{4\pi^2}\right)\left(1 - \frac{x^2}{9\pi^2}\right)\cdots$$

Where A is a constant. Since

$$\lim_{x \to 0} \frac{\sin x}{x} = 1$$

It is easy to see that $A = 1$. This approach was shown to be valid much later by Weierstrass through the **Weierstrass factorization theorem**.

> **Theorem**
>
> The Weierstrass factorization theorem asserts that every complex-valued function that is differentiable at all finite points over the whole complex plane can be represented as a product involving its zeroes. It can be seen as an extension of the fundamental theorem of algebra to complex functions.

Continuing our proof,

$$\frac{\sin x}{x} = \prod_{k=1}^{\infty} \left(1 - \frac{x^2}{(k\pi)^2}\right) \tag{9.8}$$

Substituting $x \to \pi x$ gives

$$\frac{\sin \pi x}{\pi x} = \prod_{k=1}^{\infty}\left(1 - \frac{x^2}{k^2}\right)$$

$$\therefore \sin \pi x = \pi x \prod_{k=1}^{\infty}\left(1 - \frac{x^2}{k^2}\right)$$

□

Now, back to our original proof. Notice that we can write (9.6) as

$$\Gamma(z)\Gamma(-z) = \lim_{n\to\infty} \frac{1}{z^2} \prod_{k=1}^{n}\left(\frac{z^2}{k^2} - 1\right)^{-1}$$

$$= -\frac{1}{z^2}\frac{\pi z}{\sin \pi z}$$

$$= -\frac{\pi}{z \sin \pi z}$$

Thus,

$$\Gamma(z)(-z\Gamma(-z)) = \frac{\pi}{\sin \pi z}$$

$$\implies \Gamma(z)\Gamma(1-z) = \pi \csc \pi z$$

□

9.4 Some Problems

Let us begin with a very common integral.

9.4. SOME PROBLEMS

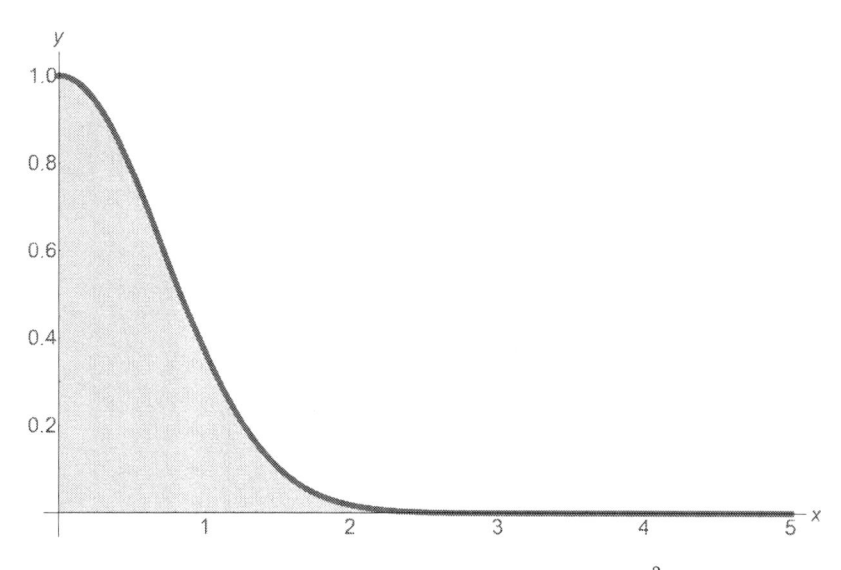

Figure 9.3: The graph of the very famous $y = e^{-x^2}$ (Gaussian integral)

Example 1: Evaluate $\displaystyle\int_0^\infty e^{-x^n} dx$ for $n > 0$.

We can substitute $u = x^n$, $dx = \frac{u^{\frac{1-n}{n}}}{n} du$ to get

$$I = \frac{1}{n}\int_0^\infty e^{-u} u^{\frac{1}{n}-1} du$$

Using the integral definition of the Gamma function gives

$$I = \frac{1}{n}\Gamma\left(\frac{1}{n}\right)$$

Since $x\Gamma(x) = \Gamma(x+1)$ we have

$$I = \Gamma\left(\frac{n+1}{n}\right) \tag{9.9}$$

Sure does seem like these special functions make a lot of things easier! Let us transition into another well-known integral.

Example 2: Evaluate $\int_{\alpha}^{\alpha+1} \ln \Gamma(x) \, dx$

Solution

Define a function

$$f(\alpha) = \int_{\alpha}^{\alpha+1} \ln \Gamma(x) \, dx$$

Taking the derivative,

$$f'(\alpha) = \ln \Gamma(\alpha + 1) - \ln \Gamma(\alpha)$$

$$= \ln \left(\frac{\Gamma(\alpha + 1)}{\Gamma(\alpha)} \right)$$

$$f'(\alpha) = \ln \alpha$$

Where in the last step we used the definition of the gamma function. Thus,

$$f(\alpha) = \int \ln \alpha \, d\alpha$$

$$= \alpha \ln \alpha - \alpha + C$$

The case when $\alpha = 0$ was evaluated back in chapter 2 (See (2.4)). This gives us the initial condition required to find the value of C.

9.4. SOME PROBLEMS

$$\Rightarrow C = \ln\left(\sqrt{2\pi}\right)$$

$$\int_\alpha^{\alpha+1} \ln\Gamma(x)\,dx = \alpha\ln\alpha - \alpha + \ln\left(\sqrt{2\pi}\right) \qquad (9.10)$$

This formula is named after the Swiss mathematician Joseph Ludwig Raabe (Raabe's formula), who derived it in 1840[2].

Example 3: Evaluate $\displaystyle\int_0^1 \Gamma\left(1+\frac{x}{2}\right)\Gamma\left(1-\frac{x}{2}\right)dx$

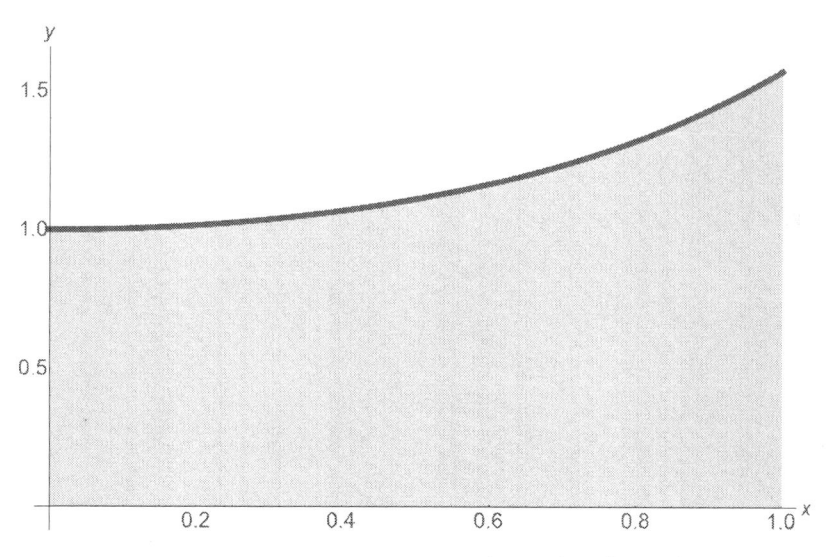

Figure 9.4: Graph of $y = \Gamma\left(1+\frac{x}{2}\right)\Gamma\left(1-\frac{x}{2}\right)$

Solution

By definition, $x\Gamma(x) = \Gamma(x+1)$,

$$I = \frac{1}{2}\int_0^1 x\Gamma\left(\frac{x}{2}\right)\Gamma\left(1-\frac{x}{2}\right)dx$$

---
[2] J. L. Raabe, *Angenäherte Bestimmung der Factorenfolge* $1 \cdot 2 \cdot 3 \cdot 4 \cdot 5 \ldots n = \Gamma(1+n) = \int x^n e^{-x}dx$, *wenn n eine sehr grosse Zahl ist*, J. Reine Angew. Math. 25 (1840), 146-159.

We can now apply Euler's reflection formula

$$\Gamma(z)\Gamma(1-z) = \frac{\pi}{\sin(\pi z)}$$

To get

$$I = \frac{1}{2}\int_0^1 \frac{\pi x}{\sin\left(\frac{\pi x}{2}\right)}\,dx$$

Substituting $u = \frac{\pi x}{2}$,

$$I = \frac{2}{\pi}\int_0^{\frac{\pi}{2}} \frac{u}{\sin u}\,du$$

Now, consider the double angle formula for $\sin x$:

$$\sin 2x = 2\sin x \cos x$$

Then,

$$\sin x = 2\sin\left(\frac{x}{2}\right)\cos\left(\frac{x}{2}\right)$$

And

$$\mathcal{I} = \int_0^{\frac{\pi}{2}} \frac{x}{\sin x}\,dx = \int_0^{\frac{\pi}{2}} \frac{x}{2\sin\left(\frac{x}{2}\right)\cos\left(\frac{x}{2}\right)}\,dx$$

$$= \int_0^{\frac{\pi}{2}} \frac{x}{2\tan\left(\frac{x}{2}\right)\cos^2\left(\frac{x}{2}\right)}\,dx$$

Substituting $x = 2\arctan u$, $dx = \frac{2}{1+u^2}du$,

$$\mathcal{I} = \int_0^1 \frac{2\arctan u}{u}\,du$$

Using the power series of arctan u,

$$\mathcal{I} = 2\int_0^1 \sum_{n=0}^\infty \frac{(-1)^n}{2n+1} u^{2n}\,du$$

By the dominated convergence theorem we can interchange the summation and integration and integrate to obtain

$$\mathcal{I} = 2\sum_{n=0}^\infty \frac{(-1)^n}{(2n+1)^2}$$

Which is equal to $2G$ by definition (See (7.1)).

$$\therefore I = \frac{4G}{\pi}$$

9.5 Exercise Problems

1. Prove that $x\Gamma(x) = \Gamma(x+1)$ using the integral definition of the gamma function.

2. Find $\int_0^1 \cos^2(\pi x)\ln\Gamma(x)\,dx$

3. Show that $\Gamma(n) = \int_0^1 \left(\ln\left(\frac{1}{x}\right)\right)^{n-1} dx$

4. Evaluate $\displaystyle\int_0^\infty \frac{x^5(e^{3x} - e^x)}{(e^x - 1)^4}\,dx$

5. Find $\displaystyle\int_0^\infty x^3 e^{-2x} \sin x \, dx$ (Hint: Use differentiation under the integral sign).

6.

> **Challenge Problem**
>
> Evaluate
> $$\int_0^1 \bigl(\ln \Gamma(x)\bigr)^2 \, dx$$
> Hint: Use the Fourier series of $\ln\bigl(\Gamma(x)\bigr)$,
>
> $$\ln \Gamma(x) - \ln\sqrt{2\pi}$$
> $$= \left(\frac{1}{\pi}\sum_{n=2}^\infty \frac{\ln n \sin(2\pi n x)}{n}\right) + \frac{(\gamma + \ln 2\pi)(1 - 2x)}{2}$$
> $$- \frac{1}{2}\ln|2\sin(\pi x)|$$

Chapter 10

Polygamma Functions

10.1 Definition

> **Definition**
>
> The polygamma function of order n, usually denoted as $\psi^{(n)}(\cdot)$, is defined as the $(n+1)^{\text{th}}$ derivative of the natural logarithm of the gamma function:
>
> $$\psi^{(n)}(z) = \frac{\mathrm{d}^{n+1}}{\mathrm{d}z^{n+1}} \ln \Gamma(z) \qquad (10.1)$$
>
> **Note**: The digamma function, $\psi^{(0)}(\cdot)$, is often expressed as $\psi(\cdot)$.

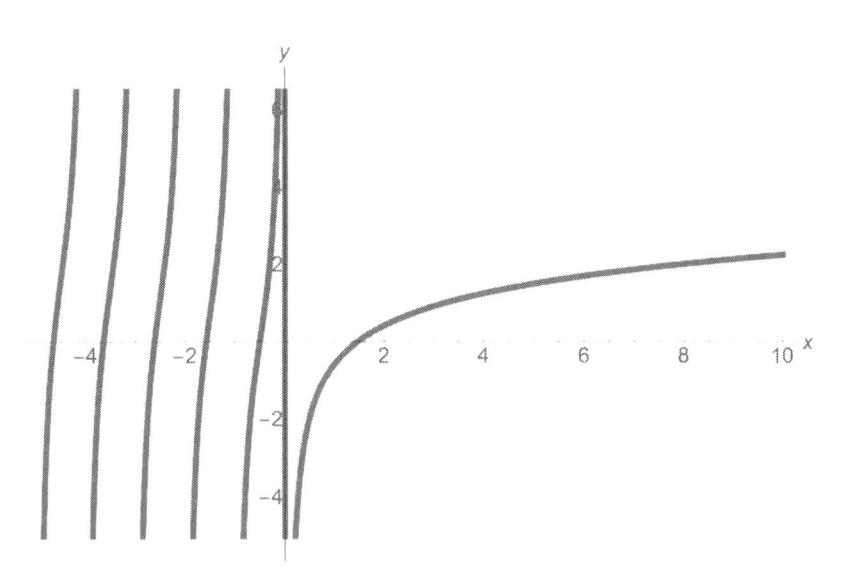

Figure 10.1: Graph of $y = \psi(x)$

10.2. SPECIAL VALUES

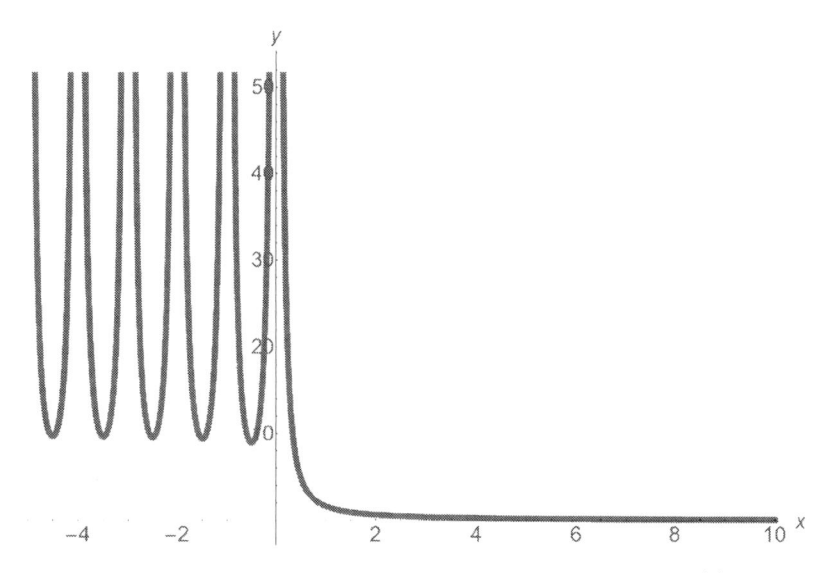

Figure 10.2: Graph of the trigamma function, $y = \psi^{(1)}(x)$

10.2 Special Values

$$\psi(1) = -\gamma$$
$$\psi\left(\frac{1}{2}\right) = -2\ln 2 - \gamma$$
$$\psi\left(\frac{1}{3}\right) = -\frac{\pi}{2\sqrt{3}} - \frac{3\ln 3}{2} - \gamma$$
$$\psi\left(\frac{1}{4}\right) = -\frac{\pi}{2} - 3\ln 2 - \gamma$$
$$\psi^{(1)}(1) = \frac{\pi^2}{6}$$
$$\psi^{(1)}\left(\frac{1}{2}\right) = \frac{\pi^2}{2}$$

10.3 Properties and Representations

The first value of the digamma function in the above section follows from (10.4), which can also be used to compute $\psi(n)$ for any $n \in \mathbb{N}$. The fractional values can be computed by **Gauss's digamma theorem**.

> **Theorem**
>
> Let n, k be positive integers with $k > n$. Then the following finite representation of the digamma function holds[a]:
>
> $$\psi\left(\frac{n}{k}\right) = -\gamma - \ln(2k) - \frac{\pi}{2}\cot\left(\frac{n\pi}{k}\right)$$
> $$+ 2\sum_{j=1}^{\lceil \frac{k}{2}\rceil - 1} \cos\left(\frac{2\pi j n}{k}\right)\ln\sin\left(\frac{\pi j}{k}\right) \qquad (10.2)$$
>
> Which follows from the digamma function's recurrence equation. Notice that if $n > k$, then the recurrence relation in (10.8) can be used.
>
> [a]Knuth, D. E. *The Art of Computer Programming, Vol. 1: Fundamental Algorithms*, 3rd ed. Reading, MA: Addison-Wesley, 1997.

We will now attempt to express the polygamma function as an infinite series.

> **Theorem**
>
> The polygamma function can be written as:
>
> $$\psi^{(k)}(z) = (-1)^{k+1}k!\sum_{n=0}^{\infty}\frac{1}{(n+z)^{k+1}} \qquad (10.3)$$

10.3. PROPERTIES AND REPRESENTATIONS

Proof. Recall the Weierstrass definition of the gamma function in (1.7):

$$\Gamma(z) = \frac{e^{-\gamma z}}{z} \prod_{k=1}^{\infty} \left(1 + \frac{z}{k}\right)^{-1} e^{z/k}$$

Taking the natural logarithm of both sides gives us

$$\ln \Gamma(z) = (-\gamma z - \ln z) + \ln \left(\prod_{k=1}^{\infty} \left(1 + \frac{z}{k}\right)^{-1} e^{z/k} \right)$$

Since $\ln \left(\prod a_n\right) = \sum \ln a_n$, we have

$$\ln \Gamma(z) = -\gamma z - \ln z + \sum_{k=1}^{\infty} \left(\frac{z}{k} - \ln\left(1 + \frac{z}{k}\right) \right)$$

Term by term differentiation is valid here since the series on the RHS converges absolutely. Thus,

$$\psi(z) = -\gamma - \frac{1}{z} + \sum_{k=1}^{\infty} \left(\frac{1}{k} - \frac{1}{z+k} \right)$$

Notice that the series above telescopes to H_z. Therefore,

$$\psi(z+1) = -\gamma + H_z \tag{10.4}$$

We can also write (10.4) as an infinite series using (6.8),

$$\psi(z+1) = -\gamma + \sum_{n=1}^{\infty} \frac{z}{n(n+z)} \tag{10.5}$$

We proceed to differentiate both sides of (10.5) to obtain

$$\psi^{(1)}(z+1) = \sum_{n=1}^{\infty} \frac{1}{(n+z)^2}$$

Substituting $z+1 \to z$ and re-indexing,

$$\psi^{(1)}(z) = \sum_{n=0}^{\infty} \frac{1}{(n+z)^2}$$

Differentiating again,

$$\psi^{(2)}(z) = -2 \sum_{n=0}^{\infty} \frac{1}{(n+z)^3}$$

In general,

$$\psi^{(k)}(z) = (-1)^{k+1} k! \sum_{n=0}^{\infty} \frac{1}{(n+z)^{k+1}}$$

Which is found by differentiating (10.5) k times.

$\square$

10.4 Some Problems

Example 1: Evaluate $\int_0^1 \frac{1-x}{1-x^3} \ln^4 x \, dx$

Solution

Define a function,

10.4. SOME PROBLEMS

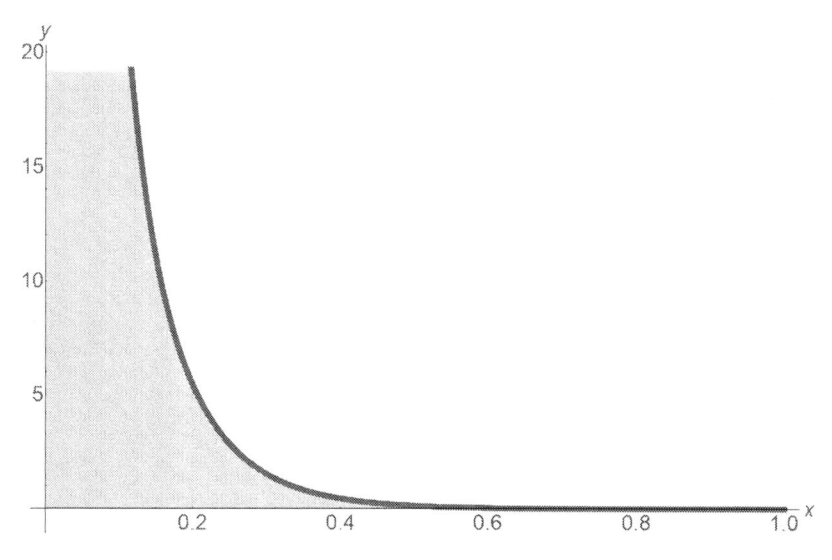

Figure 10.3: Graph of $y = \frac{1-x}{1-x^3} \ln^4 x$

$$f(\alpha) = \int_0^1 \frac{1-x}{1-x^3} x^\alpha \mathrm{d}x$$

By the series representation of $\frac{1}{1-x}$, we have:

$$\frac{1}{1-x^3} = \sum_{n=0}^\infty x^{3n}$$

And,

$$f(\alpha) = \int_0^1 \sum_{n=0}^\infty x^\alpha (1-x) x^{3n} \mathrm{d}x$$

We can then interchange summation and integration by the dominated convergence theorem

$$f(\alpha) = \sum_{n=0}^{\infty} \int_0^1 x^\alpha (1-x) x^{3n} dx$$

$$= \sum_{n=0}^{\infty} \int_0^1 x^{\alpha+3n} - x^{\alpha+3n+1} dx$$

$$= \sum_{n=0}^{\infty} \left(\frac{1}{\alpha+3n+1} - \frac{1}{\alpha+3n+2} \right) \qquad (10.6)$$

Differentiating $f(\alpha)$ under the integral sign four times gives

$$f^{(4)}(\alpha) = \int_0^1 \frac{1-x}{1-x^3} x^\alpha \ln^4 x \; dx$$

Our integral is simply $f^{(4)}(0)$. We can differentiate term-wise in (10.6) since the series is absolutely convergent. Differentiating (10.6) four times and then letting $\alpha = 0$ gives

$$I = 4! \sum_{n=0}^{\infty} \left(\frac{1}{(3n+1)^5} - \frac{1}{(3n+2)^5} \right)$$

$$= \frac{4!}{3^5} \sum_{n=0}^{\infty} \left(\frac{1}{\left(n+\frac{1}{3}\right)^5} - \frac{1}{\left(n+\frac{2}{3}\right)^5} \right)$$

By (10.3) we can write

$$I = \frac{4!}{3^5} \left(\frac{-\psi^{(4)}\left(\frac{1}{3}\right)}{4!} - \frac{-\psi^{(4)}\left(\frac{2}{3}\right)}{4!} \right)$$

10.4. SOME PROBLEMS

$$= \frac{\psi^{(4)}\left(\frac{2}{3}\right) - \psi^{(4)}\left(\frac{1}{3}\right)}{3^5}$$

Using the Mathematica call for the polygamma function, we can compute

$$I = \frac{32\pi^5}{243\sqrt{3}}$$

Example 2: Evaluate $\int_0^1 H_x \, dx$

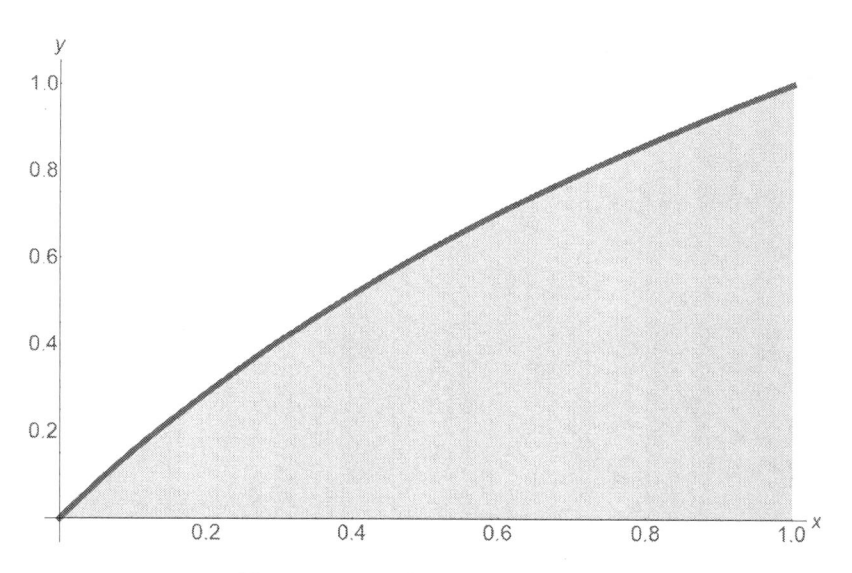

Figure 10.4: Graph of $y = H_x$

Solution

Note that we can use our result in (10.4) to get

$$H_x = \psi(x+1) + \gamma$$

CHAPTER 10. POLYGAMMA FUNCTIONS

Therefore,

$$I = \int_0^1 H_x \, dx = \int_0^1 \psi(x+1) + \gamma \, dx$$

$$= \gamma + \int_0^1 \psi(x+1) \, dx$$

By the definition of the digamma function, we have

$$I = \gamma + \bigl[\ln \Gamma(x+1)\bigr]_0^1$$

$$\therefore \int_0^1 H_x \, dx = \gamma \qquad (10.7)$$

Example 3: Evaluate $\displaystyle\int_0^1 \frac{\ln x}{x^2 + x + 1} \, dx$

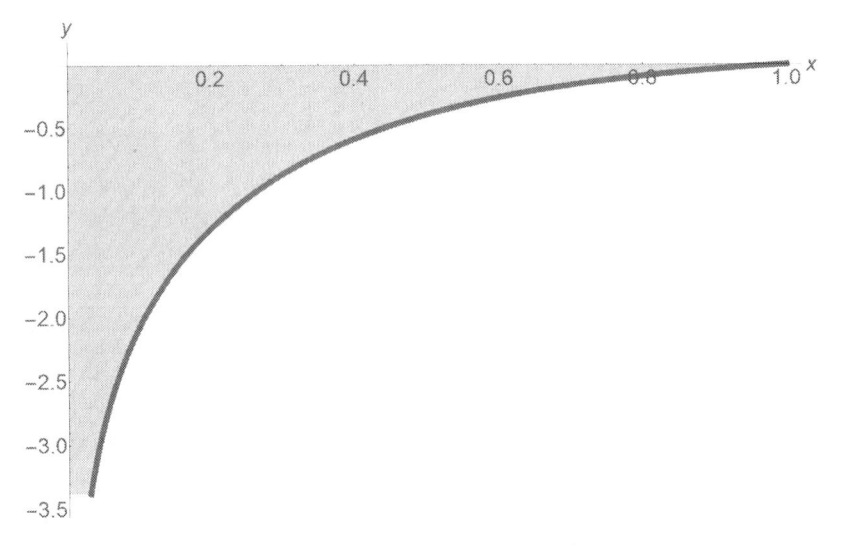

Figure 10.5: Graph of $y = \frac{\ln x}{x^2+x+1}$

10.4. SOME PROBLEMS

Solution

Recall that

$$1 - x^3 = (x^2 + x + 1)(1 - x)$$

Hence,

$$I = \int_0^1 \frac{(1-x)\ln x}{1 - x^3}\,dx$$

Splitting the integral,

$$I = \int_0^1 \frac{\ln x}{1 - x^3}\,dx - \int_0^1 \frac{x \ln x}{1 - x^3}\,dx$$

Substituting the power series of $\frac{1}{1-x^3}$ and interchanging summation and integration using the dominated convergence theorem then gives

$$\sum_{n=0}^{\infty}\left[\int_0^1 x^{3n} \ln x \, dx\right] - \sum_{n=0}^{\infty}\left[\int_0^1 x^{3n+1} \ln x \, dx\right]$$

As a standard integral, we may use:

$$\int_0^1 x^\alpha \ln x \, dx = -\frac{1}{(\alpha+1)^2}$$

Which can be easily shown by IBP. Thus,

$$I = \sum_{n=0}^{\infty} \frac{1}{(3n+2)^2} - \sum_{n=0}^{\infty} \frac{1}{(3n+1)^2}$$

Notice that the underlined sum takes the sum of squares of the form $(3n+2)^2$, and leaves out those of the form $(3n)^2$ and $(3n+1)^2$. We can then express our integral as

$$I = \zeta(2) - \sum_{n=0}^{\infty} \frac{1}{(3n+1)^2} - \sum_{n=0}^{\infty} \frac{1}{(3n)^2} - \sum_{n=0}^{\infty} \frac{1}{(3n+1)^2}$$

$$= \zeta(2) - \frac{\zeta(2)}{9} - 2\sum_{n=0}^{\infty} \frac{1}{(3n+1)^2}$$

$$= \frac{8\zeta(2)}{9} - 2\sum_{n=0}^{\infty} \frac{1}{(3n+1)^2}$$

Factoring out a 9 from the sum above we obtain:

$$I = \frac{4\pi^2}{27} - \frac{2}{9}\sum_{n=0}^{\infty} \frac{1}{\left(n+\frac{1}{3}\right)^2}$$

We can now use the series expansion of the trigamma function,

$$\psi^{(1)}(x) = \sum_{n=0}^{\infty} \frac{1}{(n+x)^2}$$

$$\therefore I = \frac{4\pi^2}{27} - \frac{2}{9}\psi^{(1)}\left(\frac{1}{3}\right)$$

Example 4: Evaluate $\displaystyle\int_0^1 \frac{5\ln x - x^5 + 1}{(1-x)\ln x}\,dx$

Solution

Denote

10.4. SOME PROBLEMS

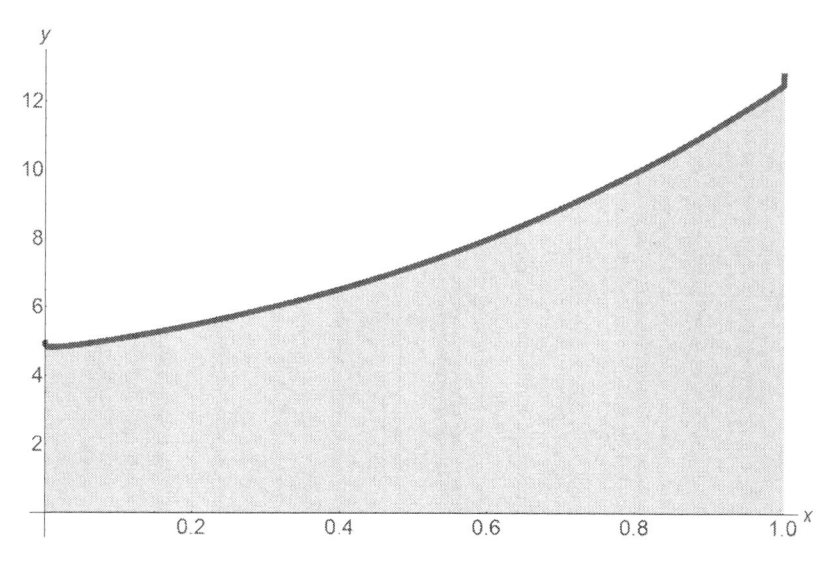

Figure 10.6: Graph of $y = \frac{5\ln x - x^5 + 1}{(1-x)\ln x}$

$$f(\alpha) = \int_0^1 \frac{\alpha \ln x - x^\alpha + 1}{\ln x (1-x)} \, \mathrm{d}x$$

Differentiating under the integral sign,

$$f'(\alpha) = \int_0^1 \frac{\ln x - x^\alpha \ln x}{(1-x)\ln x} \, \mathrm{d}x$$

$$= \int_0^1 \frac{1 - x^\alpha}{1-x} \, \mathrm{d}x$$

By (6.6), we know:

$$H_\alpha = \int_0^1 \frac{1-x^\alpha}{1-x} \, \mathrm{d}x$$

Therefore,

$$f'(\alpha) = H_\alpha$$

We can now obtain an expression for $f(\alpha)$ by integrating.

$$\begin{aligned} f(\alpha) &= \int f'(\alpha)\,d\alpha \\ &= \int H_\alpha\,d\alpha \\ &= \int \psi(\alpha+1) + \gamma\,d\alpha \\ &= \ln\Gamma(\alpha+1) + \gamma\alpha + C \end{aligned}$$

Where we used (10.4). Since $f(0) = 0$, we can write

$$C = 0$$

$$\therefore \int_0^1 \frac{\alpha \ln x - x^\alpha + 1}{\ln x (1-x)}\,dx = \ln\Gamma(\alpha+1) + \gamma\alpha$$

Plugging in $\alpha = 5$,

$$\int_0^1 \frac{5\ln x - x^5 + 1}{(1-x)\ln x}\,dx = \ln 120 + 5\gamma$$

Example 5: Evaluate $\int_0^1 x H_x\,dx$

Solution

By (10.4), we have:

$$H_x = \psi(x+1) + \gamma$$

Therefore,

10.4. SOME PROBLEMS

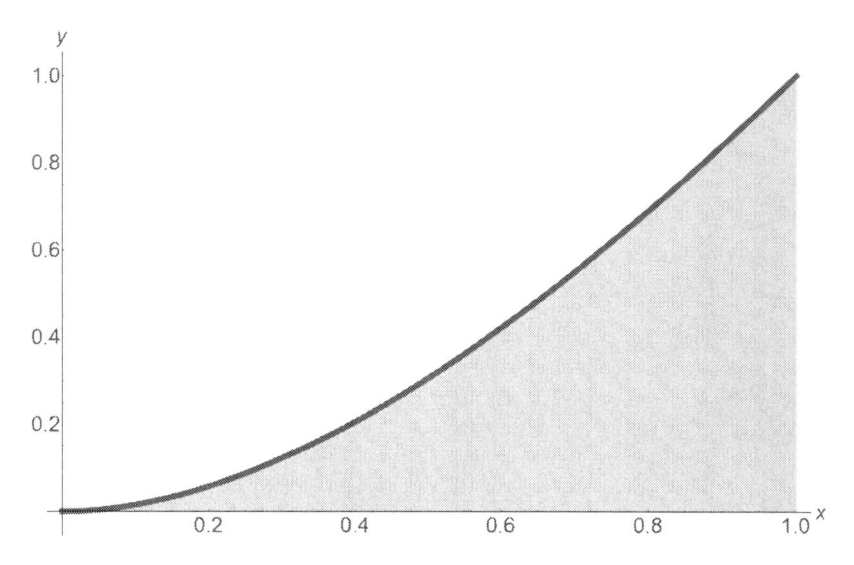

Figure 10.7: Graph of $y = xH_x$

$$I = \int_0^1 x(\psi(x+1) + \gamma)dx$$

We can use the recurrence relation

$$\psi(x+1) = \psi(x) + \frac{1}{x} \tag{10.8}$$

$$\implies I = \int_0^1 x\psi(x) + \gamma x + 1 \, dx$$

$$= \frac{\gamma}{2} + 1 + \int_0^1 x\psi(x)dx$$

We can apply IBP with $u = x$, $dv = \psi(x)dx$ on the integral above,

$$\int_0^1 x\psi(x)\mathrm{d}x = \left[x\ln\Gamma(x)\right]_0^1 - \int_0^1 \ln\Gamma(x)\,\mathrm{d}x$$

We have already evaluated this integral back in chapter 2 (See (2.4)).

$$\therefore I = \frac{\gamma}{2} + 1 - \ln\left(\sqrt{2\pi}\right)$$

Example 6: Evaluate the series $\sum_{n=0}^{\infty} \frac{1}{n^2+1}$

Solution

We will introduce the following theorem:

> **Theorem**
>
> Let x be a complex number. The following equality then holds:
> $$\sum_{k=1}^{\infty} \frac{1}{k^2+x^2} = \frac{1+\pi x \coth x\pi}{2x^2} \qquad (10.9)$$

Proof. Recall the product definition of the sine function given in (9.8):

$$\sin x = x \prod_{k=1}^{\infty} \left(1 - \frac{x^2}{(k\pi)^2}\right)$$

Substituting $x \to ix$ we get:

$$\sin ix = ix \prod_{k=1}^{\infty} \left(1 + \frac{x^2}{(k\pi)^2}\right) \qquad (10.10)$$

$$\sinh x = x \prod_{k=1}^{\infty} \left(1 + \frac{x^2}{(k\pi)^2}\right)$$

10.4. SOME PROBLEMS

Where in the last step we used the definition of the hyperbolic sine. Taking the natural logarithm of both sides,

$$\ln \sinh x = \ln x + \sum_{k=1}^{\infty} \ln\left(1 + \frac{x^2}{(k\pi)^2}\right)$$

We proceed to differentiate both sides with respect to x to obtain:

$$\frac{(\sinh x)'}{\sinh x} = \frac{1}{x} + \sum_{k=1}^{\infty} \frac{\left(1 + \frac{x^2}{(k\pi)^2}\right)'}{\left(1 + \frac{x^2}{(k\pi)^2}\right)}$$

$$\therefore \coth x = \frac{1}{x} + \sum_{k=1}^{\infty} \frac{2x}{k^2\pi^2 + x^2}$$

Here term by term differentiation is justified since the series on the RHS is absolutely convergent. Substituting $x \to \pi x$ then gives:

$$\coth \pi x = \frac{1}{\pi x} + \sum_{k=1}^{\infty} \frac{2\pi x}{k^2\pi^2 + \pi^2 x^2}$$

$$= \frac{1}{\pi x} + \frac{1}{\pi}\sum_{k=1}^{\infty} \frac{2x}{k^2 + x^2} \qquad (10.11)$$

Manipulating the above expression,

$$\sum_{k=1}^{\infty} \frac{1}{k^2 + x^2} = \frac{\pi x \coth \pi x - 1}{2x^2}$$

We can extend the sum above by adding the term when $k = 0$, which is $\frac{1}{x^2}$:

$$\sum_{k=0}^{\infty} \frac{1}{k^2 + x^2} = \frac{1 + \pi x \coth \pi x}{2x^2}$$

□

Substituting $x^2 = 1$ then gives

$$\sum_{k=0}^{\infty} \frac{1}{k^2 + 1} = \frac{1 + \pi \coth \pi}{2}$$

10.5 Exercise Problems

1. Evaluate $\lim_{x \to 0} \ln \sqrt[x]{x!}$

2. Find $\int_0^{\infty} \frac{x^2 \ln x}{e^x} \, dx$

3. Derive the reflection identity $\psi(1 - z) - \psi(z) = \pi \cot \pi z$

4. Find $\sum_{k=1}^{\infty} \frac{\psi^{(1)}(k)}{k}$

5.

> **Challenge Problem**
>
> Prove that
>
> $$\psi(x) = \int_0^{\infty} \frac{e^{-t}}{t} - \frac{e^{-xt}}{1 - e^{-t}} \, dt$$

10.5. EXERCISE PROBLEMS

6.

> **Challenge Problem**
>
> Evaluate
> $$\int_0^{\pi/2} x^3 \csc x \, \mathrm{d}x$$

Chapter 11

Beta Function

11.1 Definition

> **Definition**
>
> The beta function, or the Euler integral of the first kind, is most commonly defined as:
>
> $$\mathrm{B}(x,y) = \int_0^1 t^{x-1}(1-t)^{y-1}\mathrm{d}t \qquad (11.1)$$
>
> For $\Re(x), \Re(y) > 0$. A key property of the beta function is
>
> $$\mathrm{B}(x,y) = \frac{\Gamma(x)\Gamma(y)}{\Gamma(x+y)} \qquad (11.2)$$

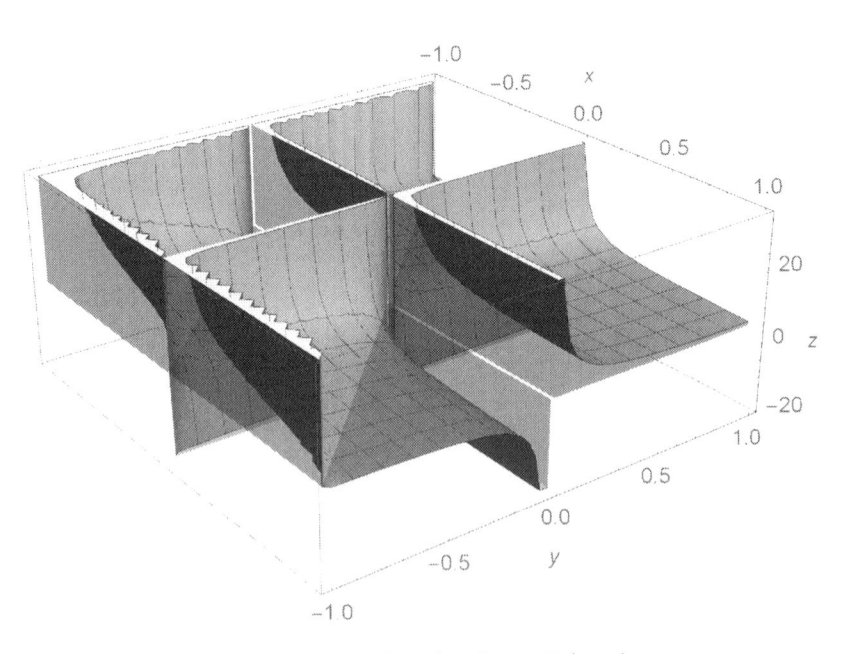

Figure 11.1: Graph of $z = \mathrm{B}(x,y)$

11.2 Special Values

$$B\left(\frac{1}{2}, \frac{1}{2}\right) = \pi$$

$$B\left(\frac{1}{3}, \frac{2}{3}\right) = \frac{2\pi}{\sqrt{3}}$$

$$B\left(\frac{1}{4}, \frac{3}{4}\right) = \pi\sqrt{2}$$

$$B(x, 1) = \frac{1}{x}$$

11.3 Properties and Representations

We will begin this section by deriving a trigonometric integral for the beta function.

> **Theorem**
>
> The beta function can be given by:
>
> $$B(x, y) = 2 \int_0^{\frac{\pi}{2}} \cos^{2x-1}(\theta) \sin^{2y-1}(\theta) d\theta, \quad \Re(x), \Re(y) > 0 \tag{11.3}$$
>
> Here, $\Re(\cdot)$ denotes the real part.

Proof. Consider the definition of the gamma function, (1.7)

$$\Gamma(n+1) = n! = \int_0^\infty t^n e^{-t} dt$$

We can then write

$$\Gamma(n+1)\Gamma(k+1) = \int_0^\infty t^n e^{-t} dt \int_0^\infty u^k e^{-u} du$$

Substituting $t = x^2$ and $u = y^2$,

$$\Gamma(n+1)\Gamma(k+1) = 4 \int_0^\infty x^{2n+1} e^{-x^2} dx \int_0^\infty y^{2k+1} e^{-y^2} dy$$

Notice that e^{-x^2} is symmetric around $x = 0$. Thus,

$$\Gamma(n+1)\Gamma(k+1) = \int_{-\infty}^\infty |x|^{2n+1} e^{-x^2} dx \int_{-\infty}^\infty |y|^{2k+1} e^{-y^2} dy \quad (11.4)$$

$$= \int_{-\infty}^\infty \int_{-\infty}^\infty |x|^{2n+1} e^{-x^2} |y|^{2k+1} e^{-y^2} dx$$

Transforming into polar coordinates with $x = r\cos\theta$, $y = r\sin\theta$

$$\Gamma(n+1)\Gamma(k+1) = \int_0^{2\pi} \int_0^\infty r \cdot e^{-r^2} |r\cos\theta|^{2n+1} |r\sin\theta|^{2k+1} dr d\theta$$

$$= \int_0^{2\pi} \int_0^\infty r^{2n+2k+3} e^{-r^2} |\cos\theta|^{2n+1} |\sin\theta|^{2k+1} dr d\theta$$

$$= \left[\int_0^\infty r^{2n+2k+3} e^{-r^2} dr \right] \left[\int_0^{2\pi} \underline{|\cos\theta|^{2n+1} |\sin\theta|^{2k+1}} d\theta \right]$$

Notice that the underlined integrand is periodic with a period of $\frac{\pi}{2}$. Thus,

$$\Gamma(n+1)\Gamma(k+1)$$
$$= 4 \left[\int_0^\infty r^{2n+2k+3} e^{-r^2} dr \right] \left[\int_0^{\frac{\pi}{2}} |\cos\theta|^{2n+1} |\sin\theta|^{2k+1} d\theta \right]$$

11.3. PROPERTIES AND REPRESENTATIONS

Since $\cos\theta$ and $\sin\theta$ are non-negative for all $\theta \in \left[0, \frac{\pi}{2}\right]$, we can get rid of the absolute value sign.

$$\Gamma(n+1)\Gamma(k+1) = 4\left[\int_0^\infty r^{2n+2k+3} e^{-r^2}\,dr\right]\left[\int_0^{\frac{\pi}{2}} \cos^{2n+1}(\theta)\sin^{2k+1}(\theta)d\theta\right]$$

Substituting $u = r^2$, $du = 2r\,dr$ gives

$$\Gamma(n+1)\Gamma(k+1) = 2\left[\int_0^\infty u^{n+k+1} e^{-u}\,du\right]\left[\int_0^{\frac{\pi}{2}} \cos^{2n+1}(\theta)\sin^{2k+1}(\theta)d\theta\right]$$

We will now use the definition of the gamma function to get

$$\Gamma(n+1)\Gamma(k+1) = 2\Gamma(n+k+2)\int_0^{\frac{\pi}{2}} \cos^{2n+1}(\theta)\sin^{2k+1}(\theta)d\theta$$

$$\frac{\Gamma(n+1)\Gamma(k+1)}{\Gamma(n+k+2)} = 2\int_0^{\frac{\pi}{2}} \cos^{2n+1}(\theta)\sin^{2k+1}(\theta)d\theta$$

By (11.2), we can express the LHS as:

$$\mathrm{B}(n+1, k+1) = 2\int_0^{\frac{\pi}{2}} \cos^{2n+1}(\theta)\sin^{2k+1}(\theta)d\theta$$

Or alternatively,

$$\mathrm{B}(x, y) = 2\int_0^{\frac{\pi}{2}} \cos^{2x-1}(\theta)\sin^{2y-1}(\theta)d\theta$$

For $\mathfrak{R}(x), \mathfrak{R}(y) > 0$. □

A shorter method to derive (11.3) is through the definition given in (11.1). The substitution $t = \cos^2 u$, $dt = -2\cos u \sin u \, du$ into (11.1) gives

$$B(x,y) = \int_0^1 t^{x-1}(1-t)^{y-1} dt = -2\int_{\frac{\pi}{2}}^0 \cos^{2x-1}(u) \sin^{2y-1}(u) \, du$$

$$= 2\int_0^{\frac{\pi}{2}} \cos^{2x-1}(u) \sin^{2y-1}(u) \, du$$

We will now prove an interesting identity.

> **Theorem**
>
> The beta function has a recursive property given by:
>
> $$B(x,y) = B(x, y+1) + B(x+1, y) \qquad (11.5)$$

Proof. Notice that by (11.2) we can write

$$B(x+1, y) = \frac{\Gamma(x+1)\Gamma(y)}{\Gamma(x+y+1)}$$

$$= \frac{x\Gamma(x)\Gamma(y)}{(x+y)\Gamma(x+y)} = \frac{x}{x+y} \cdot \frac{\Gamma(x)\Gamma(y)}{\Gamma(x+y)}$$

We now have the recurrence relation

$$B(x+1, y) = \frac{x}{x+y} \cdot B(x,y) \qquad (11.6)$$

And, by symmetry,

$$B(x, y+1) = \frac{y}{x+y} \cdot B(x,y)$$

11.3. PROPERTIES AND REPRESENTATIONS

Hence,

$$B(x, y+1) + B(x+1, y) = \frac{x}{x+y} \cdot B(x,y) + \frac{y}{x+y} \cdot B(x,y)$$

$$\therefore B(x,y) = B(x, y+1) + B(x+1, y)$$

$\square$

Now, for one more!

> **Theorem**
>
> An alternative integral representation of the beta function is given by
>
> $$B(x,y) = \int_0^\infty \frac{u^{x-1}}{(1+u)^{x+y}} du \quad , \quad \Re(x), \Re(y) > 0 \quad (11.7)$$

Proof. The substitution $t = \frac{u}{u+1}$, $dt = \frac{1}{(u+1)^2} du$ into (11.1) gives

$$B(x,y) = \int_0^1 t^{x-1}(1-t)^{y-1} dt$$

$$= \int_0^\infty \frac{u^{x-1}}{(1+u)^{x-1}(1+u)^{y-1}(1+u)^2} du$$

$$= \int_0^\infty \frac{u^{x-1}}{(1+u)^{x+y}} du$$

For $\Re(x), \Re(y) > 0$. $\square$

11.4 Some Problems

Example 1: Find $\displaystyle\int_0^\infty \frac{\ln x}{1+x^n}dx$

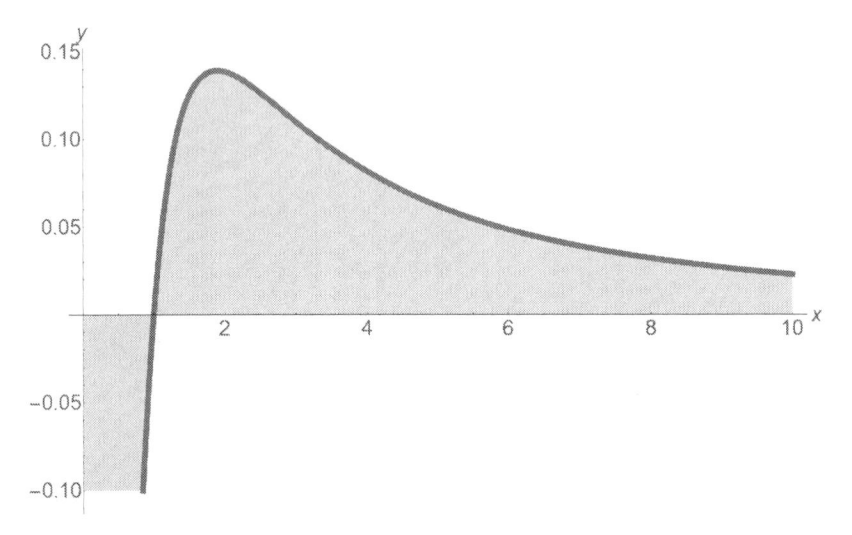

Figure 11.2: Graph of $y = \frac{\ln x}{1+x^2}$

Solution

Substituting $u = x^n$, $du = nx^{n-1}dx$ gives

$$I = \frac{1}{n^2}\int_0^\infty \frac{u^{\frac{1}{n}-1}\ln u}{1+u}du$$

From the definition of the beta function given in (11.7),

$$\mathrm{B}(x,y) = \int_0^\infty \frac{t^{x-1}}{(1+t)^{x+y}}dt$$

11.4. SOME PROBLEMS

Our integral is simply

$$I = \frac{1}{n^2} \frac{\partial}{\partial x} B(x, 1-x) \bigg|_{x=\frac{1}{n}}$$

Since

$$B(x, y) = \frac{\Gamma(x)\Gamma(y)}{\Gamma(x+y)}$$

We can express I as

$$I = \frac{1}{n^2} \frac{\partial}{\partial x} \left[\Gamma(x) \Gamma(1-x) \right]_{x=\frac{1}{n}}$$

Recall Euler's reflection formula,

$$\Gamma(x)\Gamma(1-x) = \pi \csc(\pi x)$$

Hence,

$$I = \frac{1}{n^2} \frac{\partial}{\partial x} \left[\pi \csc(\pi x) \right]_{x=\frac{1}{n}}$$
$$= -\frac{\pi^2}{n^2} \cot\left(\frac{\pi}{n}\right) \csc\left(\frac{\pi}{n}\right)$$

Example 2: Evaluate $\int_0^\infty \frac{\mathrm{d}x}{\sqrt{1+x^4}(1+x^\pi)}$

Solution
We will solve a generalized form of the integral here. Define

$$f(\alpha) = \int_0^\infty \frac{\mathrm{d}x}{\sqrt{1+x^4}(1+x^\alpha)}$$

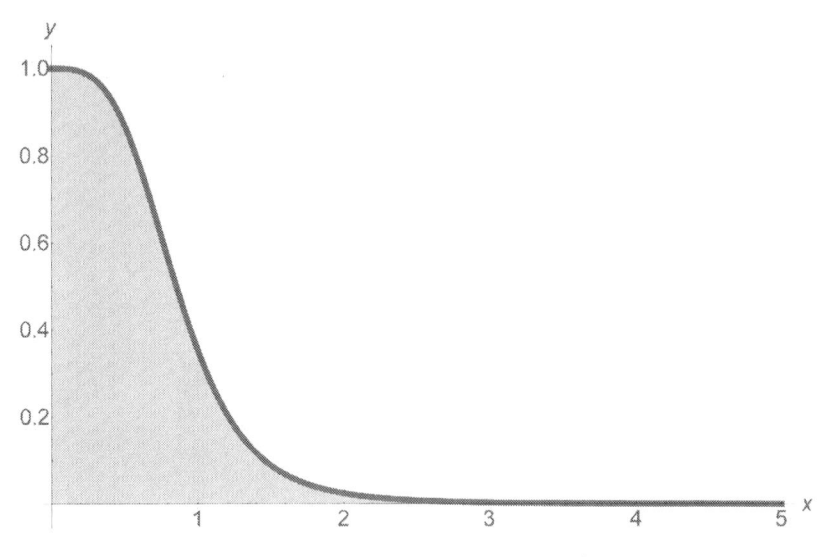

Figure 11.3: Graph of $y = \frac{1}{\sqrt{1+x^4}(1+x^\pi)}$

We then proceed to apply the u-substitution $u = \frac{1}{x}$, $\mathrm{d}x = -\frac{\mathrm{d}u}{u^2}$

$$f(\alpha) = \int_0^\infty \frac{\mathrm{d}x}{\sqrt{1+x^4}(1+x^\alpha)}$$

$$= \int_\infty^0 \frac{1}{\sqrt{1+\frac{1}{u^4}}\left(1+\frac{1}{u^\alpha}\right)} \cdot -\frac{\mathrm{d}u}{u^2}$$

$$= \int_0^\infty \frac{\mathrm{d}u}{\sqrt{u^4+1}\left(1+\frac{1}{u^\alpha}\right)}$$

$$= \int_0^\infty \frac{\mathrm{d}u}{\sqrt{u^4+1}} \cdot \frac{u^\alpha}{1+u^\alpha}$$

Therefore,

$$2f(\alpha) = \int_0^\infty \frac{\mathrm{d}u}{\sqrt{u^4+1}} \cdot \frac{u^\alpha}{1+u^\alpha} + \int_0^\infty \frac{\mathrm{d}x}{\sqrt{1+x^4}} \cdot \frac{1}{1+x^\alpha}$$

11.4. SOME PROBLEMS

After combining these integrals we obtain

$$2f(\alpha) = \int_0^\infty \frac{\mathrm{d}x}{\sqrt{1+x^4}} \cdot \underbrace{\left(\frac{x^\alpha}{1+x^\alpha} + \frac{1}{1+x^\alpha}\right)}_{=1}$$

$$2f(\alpha) = \int_0^\infty \frac{\mathrm{d}x}{\sqrt{1+x^4}}$$

Turns out that π was just a distractor! We will now substitute $x = \sqrt{\tan u}$, $\mathrm{d}x = \frac{\sec^2 u}{2\sqrt{\tan u}}\mathrm{d}u$

$$2f(\alpha) = \frac{1}{2}\int_0^{\frac{\pi}{2}} \frac{\mathrm{d}u}{\sqrt{1+\tan^2 u}} \cdot \frac{\sec^2 u}{\sqrt{\tan u}}$$

$$= \frac{1}{2}\int_0^{\frac{\pi}{2}} \frac{\mathrm{d}u}{\sec u} \cdot \frac{\sec^2 u}{\sqrt{\tan u}}$$

$$= \frac{1}{2}\int_0^{\frac{\pi}{2}} \frac{\sec u}{\sqrt{\tan u}}\mathrm{d}u$$

$$= \frac{1}{2}\int_0^{\frac{\pi}{2}} \cos^{-\frac{1}{2}}(u) \sin^{-\frac{1}{2}}(u)\mathrm{d}u$$

Using (11.3) gives

$$8f(\alpha) = \frac{\left(\Gamma\left(\frac{1}{4}\right)\right)^2}{\Gamma\left(\frac{1}{2}\right)} = \frac{\left(\Gamma\left(\frac{1}{4}\right)\right)^2}{\sqrt{\pi}}$$

Therefore,

$$\int_0^\infty \frac{\mathrm{d}x}{\sqrt{1+x^4}(1+x^\pi)} = \frac{\left(\Gamma\left(\frac{1}{4}\right)\right)^2}{8\sqrt{\pi}} \qquad (11.8)$$

Example 3: Find $\int_0^{\frac{\pi}{2}} \frac{\cos^2 x}{\sqrt{1+\cos^2 x}}\mathrm{d}x$

CHAPTER 11. BETA FUNCTION

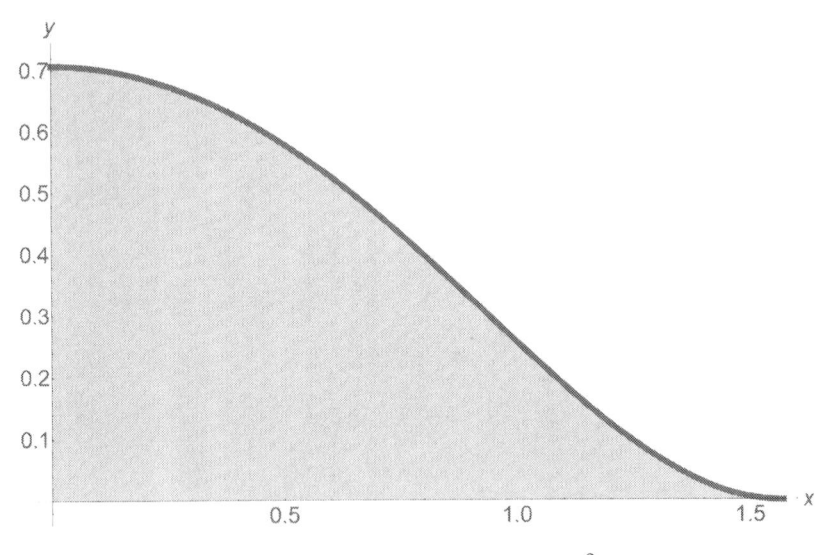

Figure 11.4: Graph of $y = \dfrac{\cos^2 x}{\sqrt{1+\cos^2 x}}$

Solution

We first substitute $u = \cos^4 x \implies du = -4\cos^3 x \sin x \, dx$. We proceed to solve for dx in terms of u.

$$dx = \frac{-du}{4\cos^3 x \sin x}$$

$$= \frac{-du}{4u^{\frac{3}{4}} \sin x}$$

Note that $\sin x = \sqrt{1 - \cos^2 u} = \sqrt{1 - \sqrt{u}}$. Therefore,

$$dx = \frac{-du}{4u^{\frac{3}{4}}\sqrt{1 - \sqrt{u}}}$$

Thus,

$$I = \int_1^0 \frac{\sqrt{u}}{\sqrt{1+\sqrt{u}}} \cdot \frac{du}{4u^{\frac{3}{4}}\sqrt{1 - \sqrt{u}}}$$

11.4. SOME PROBLEMS

$$= \frac{1}{4}\int_0^1 u^{-\frac{1}{4}}(1-u)^{-\frac{1}{2}}\,du = \frac{1}{4}B\left(\frac{3}{4},\frac{1}{2}\right)$$

We recall that
$$B(x,y) = \frac{\Gamma(x)\Gamma(y)}{\Gamma(x+y)}$$

Therefore,
$$I = \frac{\Gamma\left(\frac{3}{4}\right)\Gamma\left(\frac{1}{2}\right)}{4\Gamma\left(\frac{5}{4}\right)}$$

$$= \frac{\Gamma\left(\frac{3}{4}\right)\Gamma\left(\frac{1}{2}\right)}{\Gamma\left(\frac{1}{4}\right)}$$

Upon simplifying we obtain:

$$\therefore \int_0^{\frac{\pi}{2}} \frac{\cos^2 x}{\sqrt{1+\cos^2 x}}\,dx = \frac{\Gamma\left(\frac{3}{4}\right)^2}{\sqrt{2\pi}} \qquad (11.9)$$

Example 4: Evaluate $\displaystyle\int_0^\infty \frac{dx}{1+x^n}$

Solution

Substituting $x = \sqrt[n]{\tan^2 u}$, $dx = \frac{2}{n}\csc u \sec u \tan^{2/n} u\,du$ gives

$$I = \int_0^\infty \frac{dx}{1+x^n} = \frac{2}{n}\int_0^{\pi/2} \cos^{1-2/n} u \sin^{-1+2/n}\,du$$

Using (11.3),

$$I = \frac{1}{n}B\left(1-\frac{1}{n},\frac{1}{n}\right)$$

$$= \frac{1}{n}\Gamma\left(1-\frac{1}{n}\right)\Gamma\left(\frac{1}{n}\right)$$

$$I = \frac{\pi}{n} \csc\left(\frac{\pi}{n}\right)$$

Where in the last step we used Euler's reflection formula.

We will now prove the following result.

> **Theorem**
>
> The Legendre duplication formula is given by:
> $$\Gamma(x)\Gamma\left(x + \frac{1}{2}\right) = 2^{1-2x}\Gamma(2x)\sqrt{\pi}$$

Proof. Recall that

$$\sin 2t = 2\sin t \cos t$$
$$\implies \frac{\sin^{2x-1}(2t)}{2^{2x-1}} = \cos^{2x-1}(t)\sin^{2x-1}(t)$$

And,

$$\mathrm{B}(x,x) = 2\int_0^{\frac{\pi}{2}} \frac{\sin^{2x-1}(2t)}{2^{2x-1}} dt$$

By (11.3). The substitution $u = 2t$ then gives

$$\mathrm{B}(x,x) = 2^{1-2x}\int_0^{\pi} \sin^{2x-1}(u) du$$

Since the integrand is symmetric around $u = \frac{\pi}{2}$, we can write

$$\mathrm{B}(x,x) = 2 \cdot 2^{1-2x}\int_0^{\frac{\pi}{2}} \sin^{2x-1}(u) du$$

$$= 2^{1-2x} \cdot B\left(x, \frac{1}{2}\right)$$

$$= 2^{1-2x} \cdot \frac{\Gamma(x)\Gamma\left(\frac{1}{2}\right)}{\Gamma\left(x + \frac{1}{2}\right)}$$

Moreover,

$$B(x, x) = \frac{\Gamma(x)\Gamma(x)}{\Gamma(2x)}$$

Putting it all together,

$$\therefore \Gamma\left(x + \frac{1}{2}\right)\Gamma(x) = 2^{1-2x} \cdot \Gamma(2x)\sqrt{\pi} \qquad (11.10)$$

Since $\Gamma\left(\frac{1}{2}\right) = \sqrt{\pi}$. □

11.5 Exercise Problems

1. Prove that $B(x, y) = B(y, x)$ using the integral definition given in (11.1).

2. Evaluate $\int_0^{\pi/2} \sqrt{\sin x}\, dx$

3. Find a closed form for $\int_0^{\pi/2} \ln \sin x \ln \tan x\, dx$

4. Use differentiation under the integral sign and the beta function to show that $\int_0^1 \ln x \ln(1-x)\, dx = 2 - \frac{\pi^2}{6}$

5. Prove that the value of an integral of the form

$$\int_0^\infty \frac{\mathrm{d}x}{(1+x^a)^b (1+x^c)}$$

Is independent of c if $ab = 2$ (See (11.8)).

6. Evaluate $\displaystyle\int_0^\infty \sqrt[4]{1+x^4} - x \, \mathrm{d}x$

7.

> **Challenge Problem**
>
> Evaluate Watson's integral
>
> $$\int_0^\pi \int_0^\pi \int_0^\pi \frac{\mathrm{d}x \, \mathrm{d}y \, \mathrm{d}z}{1 - \cos x \cos y \cos z}$$
>
> Hint: Use the substitutions $x = \frac{1}{2}\tan\alpha$, $y = \frac{1}{2}\tan\beta$, and $z = \frac{1}{2}\tan\gamma$ and then use spherical coordinates.

Chapter 12

Zeta Function

12.1 Definition

Introduced by Euler in the first half of the 18th century, and then expanded on by Riemann's seminal 1859 paper *On the Number of Primes Less Than a Given Magnitude*, the zeta function continues to be vital not only in analysis but other domains of mathematics such as number theory. Take the Riemann hypothesis, a centuries old unsolved problem, which arose from investigating the nontrivial zeroes of the zeta function. This problem has intimate ties to number theory and the distribution of primes, as shown by Riemann in his 1859 paper[1].

> **Definition**
>
> The Riemann zeta function, usually denoted as $\zeta(\cdot)$, is most commonly defined as:
>
> $$\zeta(s) = \sum_{n=1}^{\infty} \frac{1}{n^s} \tag{12.1}$$

12.2 Special Values

Some notable values include:

$\zeta(-1) = -\dfrac{1}{12}$ Via analytic continuation

$\zeta(0) = -\dfrac{1}{2}$ Via analytic continuation

$\zeta(2) = \dfrac{\pi^2}{6}$

$\zeta(3) \approx 1.202$ This is known as Apery's constant

$\zeta(4) = \dfrac{\pi^4}{90}$

---

[1] B. Riemann, *Über die Anzahl der Primzahlen unter einer gegebenen Grösse*, Monatsber. Berlin. Akad. (1859).

12.2. SPECIAL VALUES

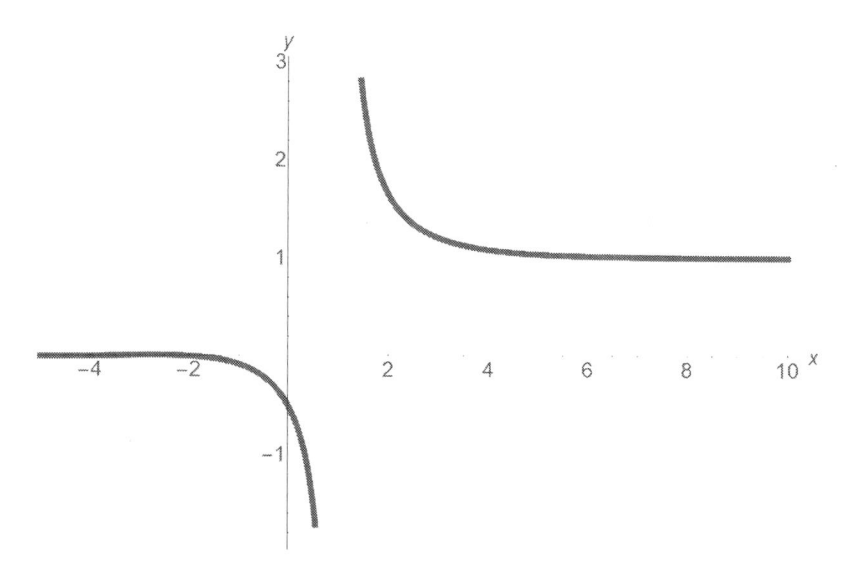

Figure 12.1: Graph of $y = \zeta(x)$. Notice that the function $\zeta(x)$ diverges at $x = 1$. Moreover, $\lim_{x \to \infty} \zeta(x) = 1$

We can easily prove some of these values. The most popular result here is the evaluation of $\zeta(2)$, which is also known as the Basel problem. We will present Euler's original approach to the problem below.

Proposition. The summation of the reciprocals of the squares of the natural numbers is given by:

$$\zeta(2) = \frac{\pi^2}{6} \tag{12.2}$$

Proof. Recall the Taylor series expansion of $\sin x$:

$$\sin x = \sum_{n=0}^{\infty} \frac{(-1)^n}{(2n+1)!} x^{2n+1} = \frac{x}{1!} - \frac{x^3}{3!} + \frac{x^5}{5!} + \cdots$$

Dividing both sides by x,

$$\frac{\sin x}{x} = \sum_{n=0}^{\infty} \frac{(-1)^n}{(2n+1)!} x^{2n} = \frac{1}{1!} - \frac{x^2}{3!} + \frac{x^4}{5!} + \cdots$$

Now, consider the infinite product definition of $\frac{\sin x}{x}$ given in (9.8).

$$\frac{\sin x}{x} = \left(1 - \frac{x^2}{\pi^2}\right)\left(1 - \frac{x^2}{4\pi^2}\right)\left(1 - \frac{x^2}{9\pi^2}\right)\cdots$$

We will now take the sum of the coefficients of x^2 (Which is allowed by Newton's Identities[2]):

$$\text{Sum of the coefficients of } x^2 = -\left(\frac{1}{\pi^2} + \frac{1}{4\pi^2} + \frac{1}{9\pi^2} + \cdots\right)$$

$$= -\frac{1}{\pi^2} \sum_{n=1}^{\infty} \frac{1}{n^2}$$

The above expression must be equal to the sum of the coefficients of x^2 in the Taylor series expansion of $\frac{\sin x}{x}$. Therefore,

$$-\frac{1}{3!} = -\frac{1}{\pi^2} \sum_{n=1}^{\infty} \frac{1}{n^2}$$

$$\therefore \sum_{n=1}^{\infty} \frac{1}{n^2} = \zeta(2) = \frac{\pi^2}{6}$$

$\square$

---

[2]See Mead, D. G. (1992). *Newtons Identities*. The American Mathematical Monthly, 99(8), 749. doi:10.2307/2324242

12.2. SPECIAL VALUES

Now, using similar logic, we will attempt to evaluate $\zeta(4)$.

Proposition. $\zeta(4)$ can be written as:

$$\zeta(4) = \frac{\pi^4}{90} \qquad (12.3)$$

Proof. Since $\zeta(2) = \frac{\pi^2}{6}$, we can write:

$$\sum_{n=1}^{\infty} \frac{1}{\pi^2 n^2} = \frac{1}{6}$$

Squaring the expression above gives

$$\left(\sum_{n=1}^{\infty} \frac{1}{\pi^2 n^2} \right)^2 = \frac{1}{36}$$

The LHS can be expressed as:

$$\left(\sum_{n=1}^{\infty} \frac{1}{\pi^2 n^2} \right)^2 = \frac{1}{\pi^4} \left(\frac{1}{1^2} + \frac{1}{2^2} + \frac{1}{3^2} \cdots \right) \left(\frac{1}{1^2} + \frac{1}{2^2} + \frac{1}{3^2} \cdots \right)$$

$$= \sum_{n=1}^{\infty} \frac{1}{\pi^4 n^4} + 2 \sum_{n<k} \frac{1}{\pi^4 n^2 k^2} \qquad (12.4)$$

To see why this is true, simply expand a product of two finite sums. For example,

$$\left(\frac{1}{1^2} + \frac{1}{2^2} + \frac{1}{3^2} \right) \left(\frac{1}{1^2} + \frac{1}{2^2} + \frac{1}{3^2} \right)$$

Similarly to how we handled the evaluation of $\zeta(2)$, we will now look at the sum of the coefficients of x^4. In the infinite product expansion of $\frac{\sin x}{x}$, this is

$$\text{Sum of coefficients of } x^4 = \frac{1}{\pi^4}\left[\frac{1}{1^2}\left(\frac{1}{2^2}+\frac{1}{3^2}+\cdots\right)\right.$$
$$\left.+\frac{1}{2^2}\left(\frac{1}{3^2}+\frac{1}{4^2}+\cdots\right)+\cdots\right]$$

$$=\sum_{n<k}\frac{1}{\pi^4 n^2 k^2}$$

This must in turn be equal to the coefficient of x^4 in the Taylor series expansion of $\frac{\sin x}{x}$. Therefore,

$$\sum_{n<k}\frac{1}{\pi^4 n^2 k^2}=\frac{1}{5!}=\frac{1}{120}$$

We can plug this result into (12.4) to obtain:

$$\frac{1}{36}=\sum_{n=1}^{\infty}\frac{1}{\pi^4 n^4}+2\left(\frac{1}{120}\right)$$

$$\sum_{n=1}^{\infty}\frac{1}{\pi^4 n^4}=\frac{1}{90}$$

$$\therefore \sum_{n=1}^{\infty}\frac{1}{n^4}=\frac{\pi^4}{90}$$

□

12.3 Properties and Representations

The most well-known form of the Riemann zeta function is its series representation, (12.1), which was considered by Euler in 1740 for positive integral values of s. The Russian mathematician Pafnuty Chebyshev then extended the function through analytic continuation to the following integral form:

$$\zeta(s) = \frac{1}{\Gamma(s)} \int_0^\infty \frac{x^{s-1}}{e^x - 1} dx \tag{12.5}$$

Perhaps the most notable aspect about the zeta function is its relation to prime numbers:

> **Theorem**
>
> The Riemann zeta function can be expressed as
>
> $$\zeta(s) = \prod_{p \text{ prime}} \frac{1}{1 - p^{-s}} \tag{12.6}$$

Euler proved the above relationship in his thesis, *Variae observationes circa series infinitas* (Various Observations about Infinite Series) in 1737. Euler's product remains an important formula in analytical number theory. It has been named the "all-important formula" by Julian Havil[3]. We will present an elementary proof of (12.6) below.

Proof. Consider the geometric series formula

$$\sum_{n=1}^\infty r^n = (1-r)^{-1}$$

---

[3] Havil, J. "The All-Important Formula." §7.1 in *Gamma: Exploring Euler's Constant*. Princeton, NJ: Princeton University Press, pp. 61-62, 2003.

For $|r| < 1$. Using this formula, we can write

$$(1 - p^{-s})^{-1} = \sum_{n=1}^{\infty} p^{-sn}$$

For some prime p and $s > 1$. The convergence of this series is easily demonstrable as the smallest prime number is 2. Notice that

$$(1 - 2^{-s})^{-1} = 1 + \frac{1}{2^s} + \frac{1}{4^s} + \frac{1}{8^s} + \cdots$$

And,

$$(1 - 3^{-s})^{-1} = 1 + \frac{1}{3^s} + \frac{1}{9^s} + \frac{1}{27^s} + \cdots$$

Multiplying the two expressions above gives

$$(1 - 2^{-s})^{-1}(1 - 3^{-s})^{-1} = \left(1 + \frac{1}{2^s} + \frac{1}{4^s} + \frac{1}{8^s} \cdots\right)$$
$$\cdot \left(1 + \frac{1}{3^s} + \frac{1}{9^s} + \frac{1}{27^s} \cdots\right)$$

Notice that this will produce all numbers of the form $2^{a_1} 3^{a_2}$, where $a_1, a_2 \in \mathbb{N}_0$. Define $S(p_1, p_2, \cdots, p_n)$ as the set of all numbers that can be expressed as $p_1^{a_1} p_2^{a_2} \cdots p_n^{a_n}$ where $a_n \in \mathbb{N}^0$. Note that there are no duplications in S by the fundamental theorem of arithmetic. Using our new notation, we have

$$(1 - 2^{-s})^{-1}(1 - 3^{-s})^{-1} = \sum_{n \in S(2,3)} \frac{1}{n^s}$$

12.3. PROPERTIES AND REPRESENTATIONS

Similarly,

$$(1-2^{-s})^{-1}(1-3^{-s})^{-1}(1-5^{-s})^{-1}\cdots(1-p_k^{-s})^{-1} = \sum_{n \in S(2,3,5,\cdots,p_k)} \frac{1}{n^s}$$

Where p_k is the kth prime number. Taking the limit as $k \to \infty$ gives:

$$\lim_{k \to \infty} (1-2^{-s})^{-1}(1-3^{-s})^{-1}(1-5^{-s})^{-1}\cdots(1-p_k^{-s})^{-1}$$

$$= \lim_{k \to \infty} \sum_{n \in S(2,3,5\cdots,p_k)} \frac{1}{n^s}$$

Since

$$\lim_{k \to \infty} S(2,3,5\cdots,p_k) \equiv \mathbb{N}$$

We can write

$$\lim_{k \to \infty} \sum_{n \in S(2,3,5\cdots,p_k)} \frac{1}{n^s} = \sum_{n=1}^{\infty} \frac{1}{n^s}$$

By the fundamental theorem of arithmetic. Thus,

$$\zeta(s) = \prod_{p \text{ prime}} \frac{1}{1 - p^{-s}}$$

□

We will now present a generating function that will aid us in our later theorem dealing with the general evaluation of any $\zeta(2n)$ where $n \in \mathbb{N}^+$.

Theorem

The generating function for even values of the zeta function is given by:

$$\frac{1}{2}(1 - \pi z \cot \pi z) = \sum_{n=1}^{\infty} \zeta(2n) z^{2n} \qquad (12.7)$$

For $|z| < 1$.

Proof. Consider the result from (9.7):

$$\sin \pi z = \pi z \prod_{k=1}^{\infty} \left(1 - \frac{z^2}{k^2}\right)$$

Taking the natural logarithm of both sides gives

$$\ln \sin \pi z = \ln \pi + \ln z + \sum_{k=1}^{\infty} \ln\left(1 - \left(\frac{z}{k}\right)^2\right)$$

By taking the derivative term by term we get (Which is justified since for $|z| < 1$, the series is absolutely convergent):

$$\pi \cot \pi z = \frac{1}{z} - \sum_{k=1}^{\infty} \frac{2z}{k^2} \cdot \frac{1}{1 - \left(\frac{z}{k}\right)^2}$$

$$= \frac{1}{z} - 2 \sum_{k=1}^{\infty} \frac{z}{k^2} \cdot \frac{1}{1 - \left(\frac{z}{k}\right)^2}$$

Now, we will use the power series expansion of $\frac{1}{1-x}$

$$\frac{1}{1-x} = \sum_{n=0}^{\infty} x^n$$

12.3. PROPERTIES AND REPRESENTATIONS

Which holds for any $|x|< 1$. Since $\left(\frac{z}{k}\right)^2 < 1$ for $|z|< 1$, we can write

$$\frac{1}{1-\left(\frac{z}{k}\right)^2} = \sum_{n=0}^{\infty}\left(\frac{z}{k}\right)^{2n}$$

And,

$$\pi \cot \pi z = \frac{1}{z} - 2\sum_{k=1}^{\infty}\sum_{n=0}^{\infty}\frac{z^{2n+1}}{k^{2n+2}}$$

$$= \frac{1}{z} - 2\sum_{n=0}^{\infty}z^{2n+1}\sum_{k=1}^{\infty}\frac{1}{k^{2n+2}}$$

By the definition of the zeta function, the inner sum is $\zeta(2n+2)$. Thus,

$$\pi \cot \pi z = \frac{1}{z} - 2\sum_{n=0}^{\infty}\zeta(2n+2)z^{2n+1}$$

We can easily rewrite the sum so it starts at $n = 1$,

$$\pi \cot \pi z = \frac{1}{z} - 2\sum_{n=1}^{\infty}\zeta(2n)z^{2n-1}$$

$$\therefore \pi z \cot \pi z = 1 - 2\sum_{n=1}^{\infty}\zeta(2n)z^{2n} \qquad (12.8)$$

After some algebraic manipulation we obtain

$$\sum_{n=1}^{\infty}\zeta(2n)z^{2n} = \frac{1}{2}\left(1 - \pi z \cot \pi z\right)$$

Using this formula, we can calculate a recursive formula for any $\zeta(2n)$, $n \in \mathbb{N}$. □

We have already seen how to evaluate $\zeta(2)$ and $\zeta(4)$. But, what about $\zeta(6)$? Maybe even $\zeta(8)$? Instead of deriving each even zeta value separately, we will present a recursive formula for evaluating any $\zeta(2n)$.

> **Theorem**
>
> The recursive formula for even zeta function values is given by:
>
> $$\zeta(2n) = \frac{(-1)^{n+1}\pi^{2n}}{(2n+1)!}n + \sum_{k=1}^{n-1}\frac{(-1)^{k+1}\pi^{2k}}{(2k+1)!}\zeta(2n-2k) \quad (12.9)$$
>
> For $n \in \mathbb{N} \setminus \{1\}$. $\setminus$ denotes the exclusion of an element from a set.

Proof. This theorem might look intimidating, but the proof we will present uses only elementary techniques. We begin by trying to evaluate a power series for $z \cot z$. Consider

$$z \cot z = \sum_{n=0}^{\infty} a(n) z^{2n}$$

Where we wrote z^{2n} instead of z^n due to the generating function we derived above. Here, $a(n)$ is an arbitrary real-valued function. Thus,

$$z \cdot \frac{\cos z}{\sin z} = \sum_{n=0}^{\infty} a(n) z^{2n}$$

12.3. PROPERTIES AND REPRESENTATIONS

$$\cos z = \frac{\sin z}{z} \sum_{n=0}^{\infty} a(n) z^{2n}$$

We can use the well-known Taylor series expansions of $\cos z$ and $\sin z$:

$$\cos z = \sum_{n=0}^{\infty} \frac{(-1)^n z^{2n}}{(2n)!}$$

$$\sin z = \sum_{n=0}^{\infty} \frac{(-1)^n z^{2n+1}}{(2n+1)!}$$

Plugging these series in gives

$$\sum_{n=0}^{\infty} \frac{(-1)^n z^{2n}}{(2n)!} = \left(\sum_{n=0}^{\infty} \frac{(-1)^n z^{2n}}{(2n+1)!} \right) \left(\sum_{n=0}^{\infty} a(n) z^{2n} \right)$$

We can apply the identity

$$\left(\sum_{n=0}^{\infty} a_n \right) \left(\sum_{n=0}^{\infty} b_n \right) = \sum_{n=0}^{\infty} \sum_{k=0}^{n} a_k b_{n-k}$$

In our case, $a_n = \frac{(-1)^n z^{2n}}{(2n+1)!}$, $b_n = a(n) z^{2n}$. Therefore,

$$\left(\sum_{n=0}^{\infty} \frac{(-1)^n z^{2n}}{(2n+1)!} \right) \left(\sum_{n=0}^{\infty} a(n) z^{2n} \right) = \sum_{n=0}^{\infty} \sum_{k=0}^{n} \frac{(-1)^k z^{2k}}{(2k+1)!} \cdot a(n-k) z^{2n-2k}$$

$$= \sum_{n=0}^{\infty} z^{2n} \left(\sum_{k=0}^{n} (-1)^k \frac{a(n-k)}{(2k+1)!} \right)$$

Then,

$$\sum_{n=0}^{\infty} \frac{(-1)^n z^{2n}}{(2n)!} = \sum_{n=0}^{\infty} \left(\sum_{k=0}^{n} (-1)^k \frac{a(n-k)}{(2k+1)!} \right) z^{2n} \qquad (12.10)$$

Notice that the coefficient of z^{2n} must be equal on both sides to preserve equality. Thus,

$$\frac{(-1)^n}{(2n)!} = \sum_{k=0}^{n} (-1)^k \frac{a(n-k)}{(2k+1)!}$$

We can isolate the first term of the series above, which is simply $a(n)$.

$$\sum_{k=0}^{n} (-1)^k \frac{a(n-k)}{(2k+1)!} = a(n) + \sum_{k=1}^{n} (-1)^k \frac{a(n-k)}{(2k+1)!}$$

We can now finally obtain an expression for $a(n)$!

$$a(n) = \frac{(-1)^n}{(2n)!} - \sum_{k=1}^{n} (-1)^k \frac{a(n-k)}{(2k+1)!} \qquad (12.11)$$

Notice that the coefficient of z^{2n} in the LHS of (12.10) when $n = 0$ is $\frac{(-1)^0}{(2\cdot 0)!} = 1$. This must be equal to the RHS's coefficient when $n = 0$. Hence,

$$1 = \sum_{k=0}^{0} (-1)^k \frac{a(n-k)}{(2k+1)!}$$

$$\therefore a(0) = 1$$

We can use this value to break up the series in (12.11) further. Isolating the value when $k = n$ gives:

12.3. PROPERTIES AND REPRESENTATIONS

$$a(n) = \frac{(-1)^n}{(2n)!} - \frac{(-1)^n}{(2n+1)!} - \sum_{k=1}^{n-1}(-1)^k \frac{a(n-k)}{(2k+1)!}$$

$$= \frac{(-1)^n 2n}{(2n+1)!} - \sum_{k=1}^{n-1}(-1)^k \frac{a(n-k)}{(2k+1)!} \qquad (12.12)$$

From (12.8) we can write

$$z \cot z = 1 - 2\sum_{n=1}^{\infty} \zeta(2n)\left(\frac{z}{\pi}\right)^{2n}$$

$$= 1 - 2\sum_{n=1}^{\infty} \zeta(2n)\frac{z^{2n}}{\pi^{2n}} \qquad (12.13)$$

By substituting $\pi z \to z$. Notice that our hypothetical series is

$$z \cot z = \sum_{n=0}^{\infty} a(n) z^{2n}$$

$$= 1 + \sum_{n=1}^{\infty} a(n) z^{2n}$$

By matching up the coefficients of our hypothetical series and the series in (12.13), we obtain that

$$a(n) = -\frac{2\zeta(2n)}{\pi^{2n}}$$

Plugging that into (12.12),

$$-\frac{2\zeta(2n)}{\pi^{2n}} = \frac{(-1)^n 2n}{(2n+1)!} - \sum_{k=1}^{n-1}(-1)^k \frac{-\frac{2\zeta(2n-2k)}{\pi^{2n-2k}}}{(2k+1)!}$$

$$\therefore \zeta(2n) = \frac{(-1)^{n+1}\pi^{2n}n}{(2n+1)!} + \sum_{k=1}^{n-1}\frac{(-1)^{k+1}\pi^{2k}}{(2k+1)!}\zeta(2n-2k)$$

Hence proved. This formula can be used to find all even zeta values:

$$\zeta(2) = \frac{\pi^2}{6}$$

$$\zeta(4) = -\frac{2\cdot\pi^4}{5!} + \frac{\pi^2}{3!}\zeta(2) = \frac{\pi^4}{90}$$

$$\zeta(6) = \frac{3\pi^6}{7!} - \frac{\pi^4}{5!}\zeta(2) + \frac{\pi^2}{3!}\zeta(4) = \frac{\pi^6}{945}$$

$$\zeta(8) = -\frac{4\cdot\pi^8}{9!} + \frac{\pi^6}{7!}\zeta(2) - \frac{\pi^4}{5!}\zeta(4) + \frac{\pi^2}{3!}\zeta(6) = \frac{\pi^8}{9450}$$

And so on. □

12.4 Some Problems

Example 1: Evaluate $\sum_{p \text{ prime}} \frac{\ln p}{p^2 - 1}$

Solution

We will start off by the prime product formula for the zeta function, (12.6):

$$\zeta(s) = \prod_{p \text{ prime}} \frac{1}{1 - p^{-s}}$$

Taking the natural logarithm of both sides,

12.4. SOME PROBLEMS

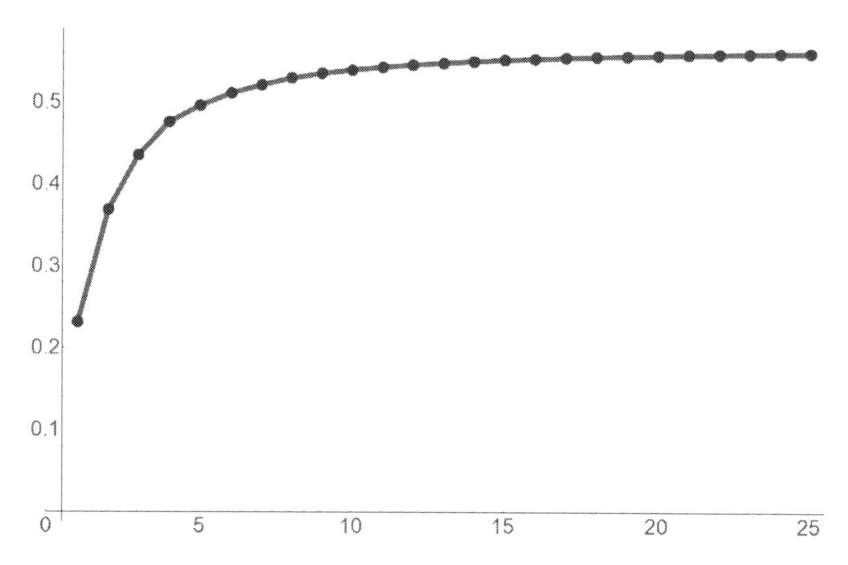

Figure 12.2: Plot of the partial sums

$$\ln(\zeta(s)) = \ln\left(\prod_{p \text{ prime}} \frac{1}{1 - p^{-s}}\right)$$

Since $\ln\left(\prod a_n\right) = \sum \ln a_n$, we have:

$$\ln(\zeta(s)) = \sum_{p \text{ prime}} \ln\left(\frac{1}{1 - p^{-s}}\right)$$

$$= -\sum_{p \text{ prime}} \ln\left(1 - p^{-s}\right)$$

We can differentiate both sides with respect to s to get:

$$-\frac{\zeta'(s)}{\zeta(s)} = \sum_{p \text{ prime}} \frac{\ln p}{p^s - 1}$$

Our desired sum is then

$$S = -\frac{\zeta'(2)}{\zeta(2)} \qquad (12.14)$$

Substituting (8.8) into (12.14) gives

$$\sum_{p \text{ prime}} \frac{\ln p}{p^2 - 1} = 12 \ln A - \gamma - \ln(2\pi)$$

Example 2: Evaluate $\displaystyle\sum_{n=2}^{\infty} \frac{\zeta(n)}{2^n}$

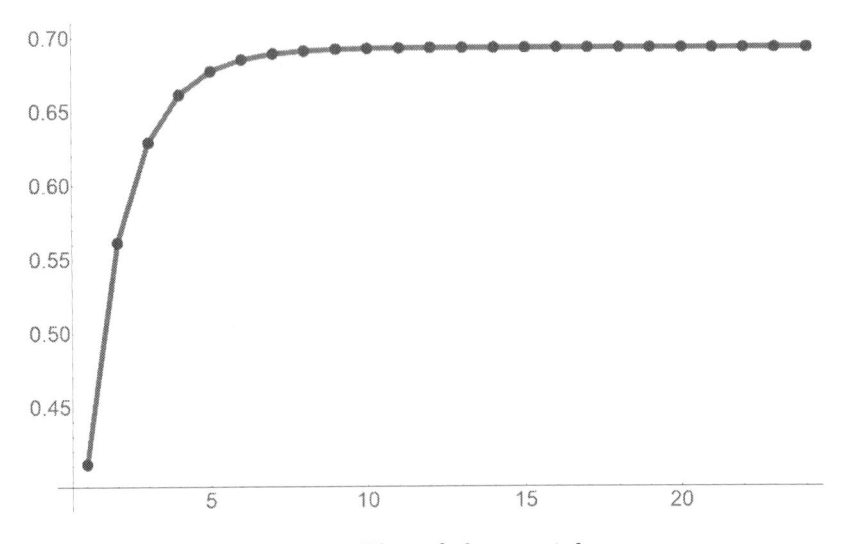

Figure 12.3: Plot of the partial sums

Solution

We will generalize this result. Define

$$f(\alpha) = \sum_{n=2}^{\infty} (-\alpha)^n \zeta(n)$$

12.4. SOME PROBLEMS

For $|\alpha| < 1$. By the definition of the zeta function, we have

$$f(\alpha) = \sum_{n=2}^{\infty} \sum_{k=1}^{\infty} \frac{(-\alpha)^n}{k^n}$$

Notice that our sum is absolutely convergent. We can then safely interchange summations to get

$$f(\alpha) = \sum_{k=1}^{\infty} \sum_{n=2}^{\infty} \frac{(-\alpha)^n}{k^n}$$

$$= \sum_{k=1}^{\infty} \sum_{n=2}^{\infty} (-1)^n \left(\frac{\alpha}{k}\right)^n$$

Recall the power series expansion of $\frac{1}{1+x}$:

$$\frac{1}{1+x} = \sum_{n=0}^{\infty} (-1)^n x^n$$

We can multiply by x^2 to obtain:

$$\frac{x^2}{1+x} = \sum_{n=2}^{\infty} (-1)^n x^n$$

Therefore,

$$f(\alpha) = \sum_{k=1}^{\infty} \left(\left(\frac{\alpha}{k}\right)^2 \cdot \frac{1}{1+\frac{\alpha}{k}} \right)$$

$$= \sum_{k=1}^{\infty} \left(\frac{\alpha^2}{k(k+\alpha)} \right)$$

$$= \alpha \sum_{k=1}^{\infty} \left(\frac{1}{k} - \frac{1}{k+\alpha} \right)$$

$$\implies f(\alpha) = \alpha \left[\gamma + \psi(\alpha+1) \right]$$

Our desired sum is $f\left(-\frac{1}{2}\right)$. Thus,

$$f\left(-\frac{1}{2}\right) = -\frac{1}{2}\left[\gamma + \psi\left(\frac{1}{2}\right)\right]$$

Since

$$\psi\left(\frac{1}{2}\right) = -2\ln 2 - \gamma$$

We have

$$\sum_{n=2}^{\infty} \frac{\zeta(n)}{2^n} = \ln 2$$

Example 4: Evaluate $\displaystyle\int_0^{\frac{\pi}{2}} \frac{\ln\cos x \ln\sin x}{\tan x} dx$

Solution

Notice that we can write

$$I = \int_0^{\frac{\pi}{2}} \frac{\ln\sin x \ln\cos x}{\tan x} dx$$

$$= \int_0^{\frac{\pi}{2}} \frac{(\ln\sin x \ln\cos x)\cos x}{\sin x} dx$$

12.4. SOME PROBLEMS

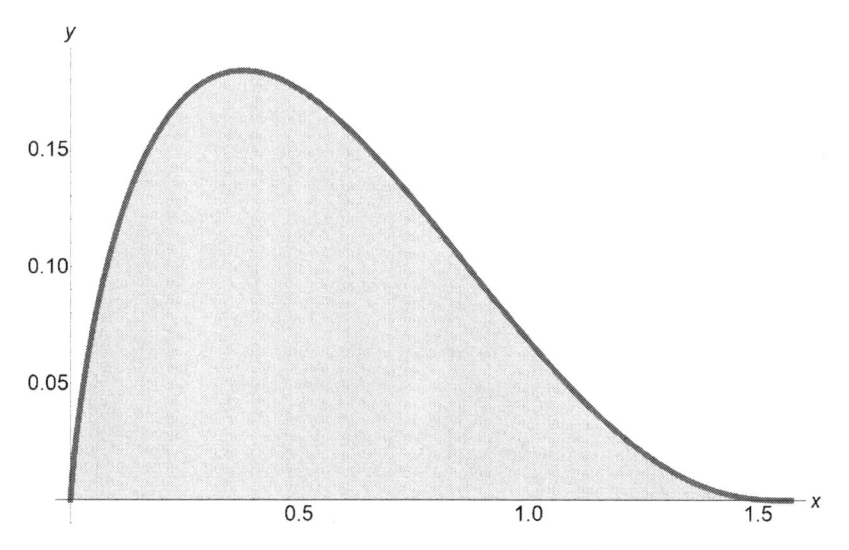

Figure 12.4: Graph of $y = \frac{\ln \cos x \ln \sin x}{\tan x}$

We now substitute $y = \sin x$, $dy = \cos x \, dx$ to get

$$I = \int_0^1 \frac{\ln y \ln\left(\sqrt{1-y^2}\right)}{y} dy$$

$$= \frac{1}{2} \int_0^1 \frac{\ln y \ln\left(1-y^2\right)}{y} dy$$

We now integrate by parts by setting $u = \ln\left(1-y^2\right)$, $dv = \frac{\ln y}{y} dy$

$$I = \frac{1}{2} \left[\frac{\ln\left(1-y^2\right) \ln^2 y}{2} \right]_0^1 + \frac{1}{2} \int_0^1 \frac{y \ln^2 y}{1-y^2} dy$$

Consider the first term. Using the power series expansion of $\ln(1-x)$, we can express the numerator as

$$A = \ln^2 y \ln\left(1-y^2\right) = -\sum_{n=1}^{\infty} \frac{y^{2n} \ln^2 y}{n}$$

It is easy to see that the expression is zero when $y = 1$. We now need to find $\lim_{y \to 0} A$.

$$\lim_{y \to 0} A = -\lim_{y \to 0} \sum_{n=1}^{\infty} \frac{y^{2n} \ln^2 y}{n} = -\sum_{n=1}^{\infty} \lim_{y \to 0} \frac{\ln^2 y}{\frac{n}{y^{2n}}}$$

We can see that for each $n \in \mathbb{N}$, there is an $\frac{\infty}{\infty}$ case here. Hence, we will use our friend: L'Hopital's rule.

$$\implies \lim_{y \to 0} A = \sum_{n=1}^{\infty} \lim_{y \to 0} \frac{\ln y}{\frac{n^2}{y^{2n}}}$$

Again, we have a $\frac{\infty}{\infty}$ case, so we will use L'Hopital's rule once more.

$$\lim_{y \to 0} A = -\sum_{n=1}^{\infty} \lim_{y \to 0} \frac{y^{2n}}{2n^3}$$

The evaluation of this sum gives

$$\lim_{y \to 0} A = 0$$

Hence,

$$\left[\frac{\ln(1 - y^2) \ln^2 y}{2} \right]_0^1 = 0 - 0 = 0$$

We then have

$$I = \frac{1}{2} \int_0^1 \frac{y \ln^2 y}{1 - y^2} dy$$

12.4. SOME PROBLEMS

Substituting $y = e^t$, $dy = e^t dt$,

$$I = \frac{1}{2}\int_{-\infty}^{0} \frac{e^{2t}t^2}{1-e^{2t}}dt$$

We then multiply by $\dfrac{e^{-2t}}{e^{-2t}}$ to get

$$I = \frac{1}{2}\int_{-\infty}^{0} \frac{t^2}{e^{-2t}-1}dt$$

Substituting $u = -2t$, $du = -2\,dt$,

$$I = \frac{1}{16}\int_{0}^{\infty} \frac{u^2}{e^u-1}du$$

By equation (12.5) we know that

$$\zeta(s) = \frac{1}{\Gamma(s)}\int_{0}^{\infty} \frac{u^{s-1}}{e^u-1}du$$

Therefore,

$$I = \frac{1}{16}\Gamma(3)\zeta(3) = \frac{\zeta(3)}{8}$$

Example 5: Evaluate $\int_{0}^{\infty} \dfrac{x^{10}}{e^{3x}-1}dx$

Rather random numbers? Must be something we have to generalize! Define

$$f(\alpha, \beta) = \int_{0}^{\infty} \frac{x^\alpha}{e^{\beta x}-1}dx$$

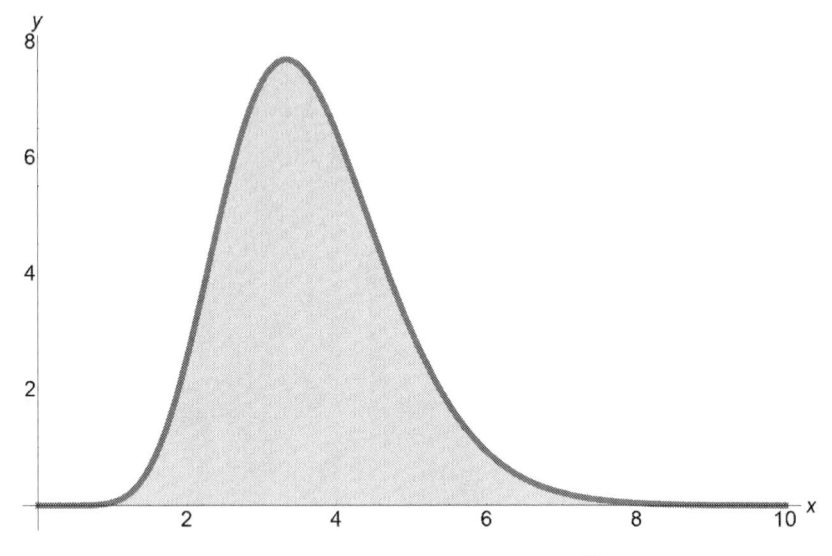

Figure 12.5: Graph of $y = \frac{x^{10}}{e^{3x}-1}$

Notice that we can multiply by $\frac{e^{-\beta x}}{e^{-\beta x}}$ to get:

$$f(\alpha, \beta) = \int_0^\infty \frac{e^{-\beta x} x^\alpha}{1 - e^{-\beta x}} \, dx$$

Using the power series expansion of $\frac{1}{1-x}$ gives

$$\frac{x}{1-x} = \sum_{n=1}^\infty x^n$$

In our desired integral, $x = e^{-\beta x}$. Thus,

$$f(\alpha, \beta) = \int_0^\infty \sum_{n=1}^\infty x^\alpha e^{-\beta x n} \, dx$$

By the dominated convergence theorem, we can interchange

summation and integration,

$$f(\alpha,\beta) = \sum_{n=1}^{\infty} \int_0^{\infty} x^{\alpha} e^{-\beta x n} \, dx$$

The substitution $u = \beta x n$, $du = \beta n \, dx$ yields:

$$f(\alpha,\beta) = \frac{1}{\beta^{\alpha+1}} \sum_{n=1}^{\infty} \frac{1}{n^{\alpha+1}} \int_0^{\infty} u^{\alpha} e^{-u} \, du$$

$$f(\alpha,\beta) = \frac{\zeta(\alpha+1)\Gamma(\alpha+1)}{\beta^{\alpha+1}}$$

Where we used the definitions of the gamma and zeta function in the last step. Our integral is then

$$f(10,4) = \frac{\zeta(11)\Gamma(11)}{3^{11}}$$

12.5 Exercise Problems

1) Evaluate $\displaystyle\sum_{n=2}^{\infty} \zeta(n) - 1$

2) Show that $\displaystyle\sum_{n=1}^{\infty} \frac{(-1)^{n+1}}{n^s} = (1 - 2^{1-s})\zeta(s)$ (Dirichlet's eta function).

3) Find $\displaystyle\int_0^{\infty} \frac{\ln x \ln\left(\frac{x}{x+1}\right)}{(1+x)^2} dx$

4) Evaluate $\displaystyle\int_0^1 \frac{x\ln^2 x}{2(1-x)^2}\,dx$

5) Find $\displaystyle\sum_{n=2}^\infty \frac{\zeta(n)-1}{n}$

6) Find the value of $\displaystyle\int_0^1 \frac{\ln(1-x)\ln(1+x)}{x}\,dx$

7) Evaluate $\displaystyle\int_0^{\pi/2} x\ln\sin x\,dx$

8) Prove that all non-trivial zeroes of the Riemann zeta function have real part $\frac{1}{2}$.

A Note on Special Functions

The chapters in this part are but an introduction to the field of special functions, but are nonetheless at the top of the utility ladder in terms of their applications. Furthermore, these functions have been investigated since the times of the great analysts such as Euler and Gauss, and have been instrumental in mathematics and science ever since.

It is worth noting that the special functions discussed in the book are what I would call *building special functions*. The more specialized functions such as the Bessel functions, spherical harmonics, Legendre polynomials, and Laguerre polynomials, etc, build upon and derive many identities from the *building* special functions. Perhaps, in a second volume, we will explore these functions!

Part IV

Applications in the Mathematical Sciences and Beyond

Chapter 13

The Big Picture

13.1 Introduction

Although the previous chapters have tremendous utility in pure mathematics, especially the domain of real and complex analysis, they are also of great utility in the mathematical sciences. The special functions discussed extensively in the previous chapters are extremely important in these fields, and arise in many advanced applications. In this part, I aim to present a refined and varied list of applications of the book's methods, techniques, and theorems.

However, we have not covered everything! The special functions we discussed in this book serve as a precursor to many more special functions in the mathematical sciences. See "A Note on Special Functions" for more.

This chapter should serve as a reminder to how beautiful results in mathematics often translate into other disciplines, and that mathematics is truly the language of the universe. This chapter also poses many interesting questions about the relationships of special functions and physical phenomena.

These intimate relationships will be explored in depth throughout this part. Many of the applications will discuss require knowledge in their respective disciplines, but the discussion of the prerequisite knowledge will be as extensive as possible to bridge that gap.

In this chapter, applications from classical mechanics to computer science will be discussed. It is worth noting that this chapter is by no means all-inclusive. The applications of the methods discussed in this book are vast, and can not be portrayed accurately in the scope of a few chapters.

13.2 Goal of the Part

The various techniques and methods, as well as the various special functions considered, are employable in a variety of standard calculus problems involving areas, arclengths, and volumes of 3-dimensional or higher objects.

However, the author would like to diverge from the *evaluate*, *find*, etc, terminology used prior in the book. This part is designed to stimulate an appreciation for not only the mathematics behind physical laws but the science as well. I hope to accomplish this through a thorough discussion of the scientific principles *and* mathematics involved.

To the best of the my knowledge, no book has ever combined such a broad and detailed list of applications with an extensive survey of integrals and series. This is done such that the reader is familiarized with the utility of the tools they have added to their toolkit, and develop a further appreciation for these tools.

Chapter 14

Classical Mechanics

14.1 Introduction

We begin this chapter by defining the **Lagrangian**.

> **Definition**
>
> The Lagrangian is the central feature of **Lagrangian mechanics**. Lagrangian mechanics uses a quantity named the Lagrangian to summarize the dynamics of the entire system. The **non relativistic Lagrangian** is defined as:
>
> $$L = \text{KE} - U$$
>
> Where U and KE represent potential energy and kinetic energy, respectively.

> **Definition**
>
> Lagrangian mechanics was introduced by the Italian-French mathematician and astronomer Joseph-Louis Lagrange in 1788. It introduced a more systematic and sophisticated way of analyzing classical systems compared to Newtonian mechanics. No new physics is introduced in Lagrangian mechanics, however, the focus is shifted from analyzing forces to energies.

14.1.1 The Lagrange Equations

When using the Lagrangian, either the Lagrange equations of the first kind or the Lagrange equations of the second kind are used to analyze a particular system.

The Lagrange equations for any general coordinate q_i are:[1]

---
[1] Hand, L. N.; Finch, J. D. *Analytical Mechanics (2nd ed.)*. Cambridge

14.2. THE FALLING CHAIN

- First kind

$$\frac{\partial L}{\partial \mathbf{r}_n} - \frac{d}{dt}\frac{\partial L}{\partial \dot{\mathbf{r}}_n} + \sum_{n=1}^{N} \lambda_k \frac{\partial f_i}{\partial \mathbf{r}_n} = 0 \qquad (14.1)$$

- Second kind

$$\frac{d}{dt}\left(\frac{\partial L}{\partial \dot{q}_n}\right) = \frac{\partial L}{\partial q_n} \qquad (14.2)$$

Where the subscript n denotes the nth particle, N denotes the number of constraints and λ_k denotes the **Lagrange multiplier** for the kth constraint equation. Also,

$$\frac{\partial}{\partial \mathbf{r}_k} \equiv \left(\frac{\partial}{\partial x_k}, \frac{\partial}{\partial y_k}, \frac{\partial}{\partial z_k}\right), \quad \frac{\partial}{\partial \dot{\mathbf{r}}_k} \equiv \left(\frac{\partial}{\partial \dot{x}_k}, \frac{\partial}{\partial \dot{y}_k}, \frac{\partial}{\partial \dot{z}_k}\right)$$

Where $\mathbf{r} = (x, y, z)$ denotes a particle's position in space and an overhead dot denotes a time derivative.

14.2 The Falling Chain

Find the time it takes for a heavy chain of mass M and length L to fall into complete extension if initially the chain is at rest and point B is held right next to A and then released. There are no internal friction forces so mechanical energy is conserved. Assume the chain is inextensible and perfectly flexible.

Solution

Define our 1-dimensional axis as beginning from the ceiling, with the + direction pointing down. Hence, our initial potential energy is equal to 0.

University Press. p. 23. ISBN 9780521575720.

CHAPTER 14. CLASSICAL MECHANICS

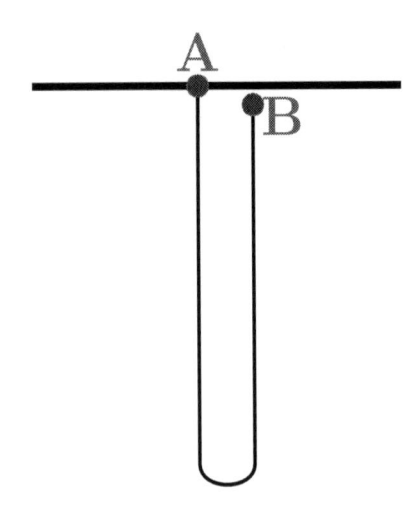

Figure 14.1: Visualization of the falling chain. Figure generated using TikZ software

When point B is x away from the ceiling, or at point x, the LHS of the chain is $\frac{L+x}{2}$ in length. Given that the density of the chain is $\frac{M}{L}$, we can write the LHS's mass as

$$M_{\text{LHS}} = \frac{L+x}{2L}M$$

Also, the LHS has a center of mass located at

$$\text{CM}_{LHS} = \frac{L+x}{4}$$

For the RHS, notice that it has length $\frac{L-x}{2}$ when B is at x, and has a mass

$$M_{\text{RHS}} = \frac{L-x}{2L}M$$

The LHS of the chain begins at x, so its center of mass is at

$$\text{CM}_{RHS} = x + \frac{L-x}{4} = \frac{L+3x}{4}$$

14.2. THE FALLING CHAIN

The total gravitational potential energy of our chain system is simply the sum of the potential energies of both sides. Equivalently,
$$U = U_{\text{LHS}} + U_{\text{RHS}}$$
Where
$$U_{\text{LHS}} = -g \cdot M_{\text{LHS}} \cdot \text{CM}_{LHS}$$
$$U_{\text{RHS}} = -g \cdot M_{\text{RHS}} \cdot \text{CM}_{RHS}$$
Here, g denotes the local acceleration due to gravity. We can then write

$$U = -\frac{Mg}{4L}\left(L^2 + 2Lx - x^2\right)$$

Notice that the LHS is stationary, but the RHS is in motion. The kinetic energy is then:

$$\text{KE}_{\text{LHS}} + \text{KE}_{\text{LHS}} = 0 + \frac{\dot{x}^2}{2}M_{\text{RHS}}$$
$$= \frac{M(L-x)\dot{x}^2}{4L}$$

Where $\dot{x}$ is simply the instantaneous velocity of the RHS when point B is at point x. Notice that the overall energy of the system must be constant by the principle of the conversation of energy, i.e.
$$E_x = E_{\text{initial}}$$
$$U_x + \text{KE}_x = U_{\text{initial}} + \cancelto{0}{\text{KE}_{\text{initial}}}$$

The initial energy of the system had no kinetic component, as we are considering the system before free fall. The center of mass of the initial system is at $\frac{L}{4}$. Hence,

$$\frac{M(L-x)\dot{x}^2}{4L} - \frac{Mg}{4L}\left(L^2 + 2Lx - x^2\right) = -\frac{MLg}{4}$$

We want to isolate $\dot{x}$ so that we can figure out the time it takes for B to reach $x = L$. With some algebraic manipulations we get

$$(L-x)\dot{x}^2 - g\left(L^2 + 2Lx - x^2\right) = -gL^2$$

$$(L-x)\dot{x}^2 = gL^2 + 2gLx - gx^2 - gL^2$$

$$(L-x)\dot{x}^2 = 2Lgx - gx^2 = gx(2L-x)$$

$$\dot{x} = \sqrt{\frac{gx(2L-x)}{L-x}}$$

We can solve for $\mathrm{d}t$, the infinitesmal change in time, as:

$$\frac{\mathrm{d}t}{\mathrm{d}x} = \sqrt{\frac{L-x}{gx(2L-x)}}$$

$$\mathrm{d}t = \sqrt{\frac{L-x}{gx(2L-x)}}\,\mathrm{d}x$$

Therefore, the time it takes for point B to reach $x = L$ from $x = 0$ is simply the sum of the infinitesimal $\mathrm{d}t$'s as $\mathrm{d}t \to 0$. We have to sum those infinitesimal time intervals from $x = 0$, our initial condition, to $x = L$, the fully extended position. From chapter two, we know this notion as the definite integral. Taking the integral from $x = 0$ to $x = L$ of both sides we have

$$T = \int_0^L \sqrt{\frac{L-x}{gx(2L-x)}}\,\mathrm{d}x \qquad (14.3)$$

Substituting $x = 2L\sin^2\left(\frac{u}{2}\right)$, $\mathrm{d}x = 2L\sin\left(\frac{u}{2}\right)\cos\left(\frac{u}{2}\right)\mathrm{d}u = L\sin u$ gives,

14.2. THE FALLING CHAIN

$$T = \int_0^{\frac{\pi}{2}} \sqrt{\frac{L\left(1 - 2\sin^2\left(\frac{u}{2}\right)\right)}{2gL\sin^2\left(\frac{u}{2}\right)\left(2L - 2L\sin^2\left(\frac{u}{2}\right)\right)}} \cdot L\sin u \, du$$

Notice that

$$1 - 2\sin^2\left(\frac{u}{2}\right) = 1 - 2\left(\sqrt{\frac{1 - \cos u}{2}}\right)^2$$
$$= 1 - 1 + \cos u$$
$$= \cos u$$

Hence,

$$T = L\int_0^{\frac{\pi}{2}} \sqrt{\frac{\cos u}{4gL\sin^2\left(\frac{u}{2}\right)\left(1 - \sin^2\left(\frac{u}{2}\right)\right)}} \cdot \sin u \, du$$

$$= L\int_0^{\frac{\pi}{2}} \sqrt{\frac{\cos u \sin^2 u}{4gL\sin^2\left(\frac{u}{2}\right)\cos^2\left(\frac{u}{2}\right)}} du$$

Simplifying,

$$T = \sqrt{\frac{L}{g}} \int_0^{\frac{\pi}{2}} \sqrt{\cos u} \, du$$

Now we have a form we can easily evaluate analytically! Using the trigonometric form of the beta function, we can express T as

$$T = \sqrt{\frac{L}{g}} \cdot \frac{1}{2}\text{B}\left(\frac{3}{4}, \frac{1}{2}\right)$$

$$= \sqrt{\frac{L}{g}} \cdot \frac{\Gamma\left(\frac{3}{4}\right)\Gamma\left(\frac{1}{2}\right)}{2\Gamma\left(\frac{5}{4}\right)}$$

Upon simplifying we get

$$T = \left(\Gamma\left(\frac{3}{4}\right)\right)^2 \sqrt{\frac{2L}{g\pi}}$$

Alternate Solution

The substitution $x = L\sin^2 u$, $\mathrm{d}x = 2L\sin u \cos u\, \mathrm{d}u$ into (14.3) gives

$$T = \int_0^L \sqrt{\frac{L-x}{gx(2L-x)}}\, \mathrm{d}x$$

$$= \int_0^{\pi/2} \sqrt{\frac{L - L\sin^2 u}{gL\sin^2 u(2L - L\sin^2 u)}} \cdot 2L\sin u \cos u\, \mathrm{d}u$$

$$= 2\sqrt{\frac{L}{g}} \int_0^{\pi/2} \frac{\cos^2 u}{\sqrt{2 - \sin^2 u}}\, \mathrm{d}u$$

Using the identity $\cos^2 u = 1 - \sin^2 u$ then gives

$$T = 2\sqrt{\frac{L}{g}} \int_0^{\pi/2} \frac{\cos^2 u}{\sqrt{1 + \cos^2 u}}\, \mathrm{d}u$$

We have already evaluated this integral in (11.9)! Plugging our result from there gives the same result,

$$T = \left(\Gamma\left(\frac{3}{4}\right)\right)^2 \sqrt{\frac{2L}{g\pi}}$$

14.3 The Pendulum

A simple pendulum consists of a mass m suspended by a string of length L such that it can swing in a plane, as shown in figure 14.2. Find the period of the pendulum for 180° swings.

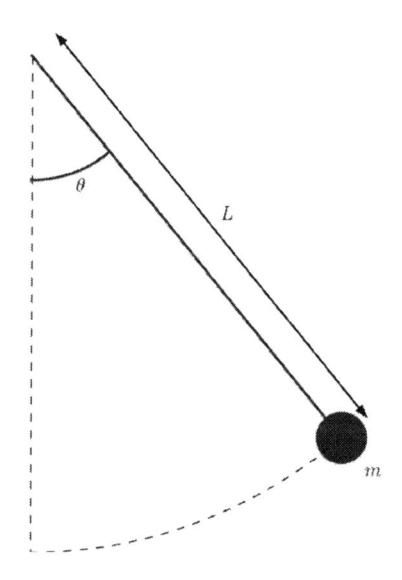

Figure 14.2: In here, $\theta_{\text{initial}} = 90$. This figure was generated through the software Pysketcher

Solution

The kinetic energy of the pendulum is:

$$\text{KE} = \frac{m\left(L\dot{\theta}\right)^2}{2}$$

Since for circular motion, $v_{\text{tangential}} = L\dot{\theta}$. Here $\dot{\theta}$ denotes the angular velocity, or the derivative of angular position.

If we define the potential energy to be $U = 0$ when the mass is at its lowest point, we can write

$$U = -mLg\cos\theta$$

When the string is at θ degrees from parallel. Now, we will calculate the *Lagrangian*.

$$\mathcal{L} = \text{KE} - U$$

$$= mgL\cos\theta + \frac{m\left(L\dot\theta\right)^2}{2}$$

We will now use Lagrange's equation of the second kind, (14.2). Note that

$$\frac{\partial \mathcal{L}}{\partial \theta} = -mgL\sin\theta$$

And,

$$\frac{\partial \mathcal{L}}{\partial \dot\theta} = mL^2\dot\theta$$

So,

$$\frac{\mathrm{d}}{\mathrm{d}t}\left(\frac{\partial L}{\partial \dot\theta}\right) = mL^2\ddot\theta$$

Plugging our values into (14.2),

$$\frac{\mathrm{d}}{\mathrm{d}t}\left(\frac{\partial \mathcal{L}}{\partial \dot q_n}\right) = \frac{\partial \mathcal{L}}{\partial q_n}$$

$$\implies \frac{\mathrm{d}}{\mathrm{d}t}\left(\frac{\partial \mathcal{L}}{\partial \dot\theta}\right) = \frac{\partial \mathcal{L}}{\partial \theta}$$

14.3. THE PENDULUM

Solving for $\ddot\theta$,
$$\therefore mL\ddot\theta = -mgL^2\sin\theta$$

$$\ddot\theta = -\frac{g\sin\theta}{L} \tag{14.4}$$

Note that we can not approximate $\sin\theta \approx \theta$, since we are dealing with 180° swings.

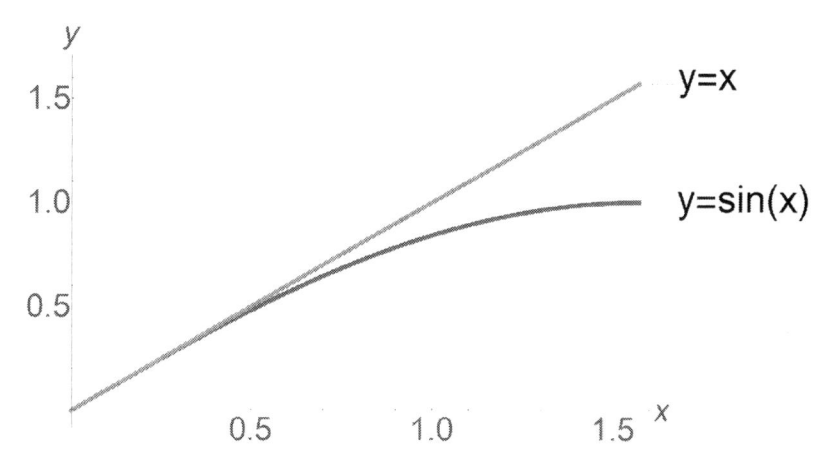

Figure 14.3: Notice that for large x, this approximation is not valid

We now want to obtain an expression for $\dot\theta$. Consider multiplying (14.4) by $\dot\theta$,

$$\dot\theta\ddot\theta = -\frac{g\sin\theta}{L}\dot\theta$$

$$\therefore \dot\theta \, d\dot\theta = -\frac{g\sin\theta}{L}\, d\theta$$

Integrating both sides,

$$\frac{\dot{\theta}^2}{2} = \frac{g\cos\theta}{L} + C$$

Since we are considering 180° swings, the pendulum is stationary at $\theta = 90°$. Hence, $\dot{\theta} = 0$ when $\theta = \frac{\pi}{2}$ and

$$C = 0$$
$$\implies \frac{\dot{\theta}^2}{2} = \frac{g\cos\theta}{L}$$

$$\dot{\theta} = \sqrt{\frac{2g\cos\theta}{L}}$$

Since $\dot{\theta} = \frac{d\theta}{dt}$, we can separate θ and t as:

$$\frac{d\theta}{dt} = \sqrt{\frac{2g\cos\theta}{L}}$$

$$\therefore \frac{d\theta}{\sqrt{\cos\theta}} = \sqrt{\frac{2g}{L}}\,dt$$

Note that the interval $\theta = 0$ to $\theta = \frac{\pi}{2}$ corresponds with the interval $t = 0$ to $t = \frac{T}{4}$, where T denotes the period. Thus,

$$\int_0^{\frac{\pi}{2}} \frac{d\theta}{\sqrt{\cos\theta}} = \int_0^{\frac{T}{4}} \sqrt{\frac{2g}{L}}\,dt = \frac{T\sqrt{\frac{2g}{L}}}{4}$$

Solving for T,

$$T = 4\sqrt{\frac{L}{2g}} \int_0^{\frac{\pi}{2}} \frac{d\theta}{\sqrt{\cos\theta}}$$

14.4. POINT MASS IN A FORCE FIELD

Using (11.3),

$$B(x,y) = 2\int_0^{\frac{\pi}{2}} \cos^{2x-1}(\theta)\sin^{2y-1}(\theta)d\theta \ , \ \Re(x), \Re(y) > 0$$

We have

$$T = 2\sqrt{\frac{L}{2g}}B\left(\frac{1}{4},\frac{1}{2}\right)$$

We can convert this to a product of gamma function values, as shown in (11.2).

$$T = 2\sqrt{\frac{L}{2g}} \cdot \frac{\Gamma\left(\frac{1}{4}\right)\Gamma\left(\frac{1}{2}\right)}{\Gamma\left(\frac{3}{4}\right)}$$

$$= \sqrt{\frac{L}{\pi g}} \cdot \Gamma^2\left(\frac{1}{4}\right)$$

14.4 Point Mass in a Force Field

A point mass of mass m was released from rest at $x = x_0$ under the influence of a force field of magnitude $F = \frac{k}{x^{n+1}}$ for $k, n \in \mathbb{R}^+$ directed towards the origin. How long would it take for the point mass to reach the origin?

Solution

Note that the case $n = 1$ corresponds with the *inverse square law* of gravity. However, we want to generalize this problem

CHAPTER 14. CLASSICAL MECHANICS

to a force of arbitrary order. Consider Newton's second law of motion

$$F = ma = m\frac{\mathrm{d}^2 x}{\mathrm{d}t^2} \qquad (14.5)$$

Denoting velocity by v, we have

$$\frac{\mathrm{d}^2 x}{\mathrm{d}t^2} = \frac{\mathrm{d}v}{\mathrm{d}t}$$

Therefore, we can write (14.5) as

$$F = m\left(\frac{\mathrm{d}v}{\mathrm{d}x}\right)\left(\frac{\mathrm{d}x}{\mathrm{d}t}\right) = mv\frac{\mathrm{d}v}{\mathrm{d}x}$$

By the chain rule. Since the only force acting on our point mass is due to the force field, we begin with

$$-\frac{k}{x^{n+1}} = mv\frac{\mathrm{d}v}{\mathrm{d}x}$$

Rewriting the expression above and integrating both sides,

$$-\frac{k}{mx^{n+1}}\,\mathrm{d}x = v\,\mathrm{d}v$$

$$\int -\frac{k}{mx^{n+1}}\,\mathrm{d}x = \int v\,\mathrm{d}v$$

$$\frac{k}{mn}x^{-n} + C = \frac{v^2}{2}$$

$$v^2 = \frac{2k}{mn}x^{-n} + C$$

14.4. POINT MASS IN A FORCE FIELD

Since our point mass starts at rest, we know that $v = 0$ when $x = x_0$. Therefore,

$$C = -\frac{2k}{mnx_0^n}$$

And,

$$v^2 = \frac{2k}{mn}x^{-n} - \frac{2k}{mnx_0^n}$$

$$= \frac{2k}{mn}\left(\frac{1}{x^n} - \frac{1}{x_0^n}\right) \tag{14.6}$$

Expressing this in derivative notation,

$$\left(\frac{dx}{dt}\right)^2 = \frac{2k}{mn}\left(\frac{1}{x^n} - \frac{1}{x_0^n}\right)$$

We now proceed to solve for dt so we can integrate to obtain time

$$\frac{dx}{dt} = \pm\sqrt{\frac{2k}{mn}\left(\frac{1}{x^n} - \frac{1}{x_0^n}\right)}$$

$$\frac{dt}{dx} = \pm\sqrt{\frac{mn}{2k}} \cdot \frac{1}{\sqrt{\frac{1}{x^n} - \frac{1}{x_0^n}}}$$

$$\therefore dt = \pm\sqrt{\frac{mn}{2k}} \cdot \frac{1}{\sqrt{\frac{1}{x^n} - \frac{1}{x_0^n}}}dx \tag{14.7}$$

Notice that

$$\frac{1}{\sqrt{\frac{1}{x^n}-\frac{1}{x_0^n}}} = \frac{1}{\sqrt{\frac{x_0^n - x^n}{x_0^n x^n}}}$$

$$= \frac{x_0^{n/2}}{\sqrt{\frac{x_0^n - x^n}{x^n}}}$$

$$= \frac{x_0^{n/2}}{\sqrt{\left(\frac{x_0}{x}\right)^n - 1}}$$

Therefore, (14.7) can be rewritten as

$$\mathrm{d}t = \pm\sqrt{\frac{mn}{2k}} \cdot \frac{x_0^{n/2}}{\sqrt{\left(\frac{x_0}{x}\right)^n - 1}} \mathrm{d}x$$

Integrating from 0 to T,

$$\int_0^T \mathrm{d}t = \pm x_0^{n/2}\sqrt{\frac{mn}{2k}} \int_{x_0}^0 \frac{1}{\sqrt{\left(\frac{x_0}{x}\right)^n - 1}} \mathrm{d}x$$

The substitution $u = \frac{x}{x_0}$, $\mathrm{d}x = x_0 \, \mathrm{d}u$ then gives

$$T = \pm x_0^{n/2}\sqrt{\frac{mn}{2k}} \int_1^0 \frac{x_0}{\sqrt{\frac{1}{u^n} - 1}} \mathrm{d}u$$

$$= \pm x_0^{1+\frac{n}{2}}\sqrt{\frac{mn}{2k}} \int_1^0 \frac{1}{\sqrt{\frac{1}{u^n} - 1}} \mathrm{d}u$$

Since T is necessarily positive, we can eliminate the $\pm$ symbol.

14.4. POINT MASS IN A FORCE FIELD

$$T = x_0^{1+\frac{n}{2}}\sqrt{\frac{mn}{2k}} \int_0^1 \frac{1}{\sqrt{\frac{1}{u^n}-1}} du$$

$$T = x_0^{1+\frac{n}{2}}\sqrt{\frac{mn}{2k}} \int_0^1 \frac{1}{\sqrt{\frac{1}{u^n}(1-u^n)}} du$$

$$= x_0^{1+\frac{n}{2}}\sqrt{\frac{mn}{2k}} \int_0^1 \frac{u^{n/2}}{\sqrt{1-u^n}} du$$

Substituting $t = u^n$, $du = \frac{t^{\frac{1}{n}-1}}{n} dt$,

$$T = x_0^{1+\frac{n}{2}}\sqrt{\frac{mn}{2k}} \int_0^1 \frac{\sqrt{t}}{\sqrt{1-t}} \cdot \left(\frac{t^{\frac{1}{n}-1}}{n}\right) dt$$

$$= x_0^{1+\frac{n}{2}}\sqrt{\frac{m}{2nk}} \int_0^1 \frac{t^{\frac{1}{n}-\frac{1}{2}}}{\sqrt{1-t}} dt$$

$$= x_0^{1+\frac{n}{2}}\sqrt{\frac{m}{2nk}} \int_0^1 t^{\frac{1}{n}-\frac{1}{2}} (1-t)^{-\frac{1}{2}} dt \qquad (14.8)$$

The reader should recognize the integral above as the main form of the beta function, defined in (11.1). Using the beta-gamma relationship then gives

$$\int_0^1 t^{\frac{1}{n}-\frac{1}{2}} (1-t)^{-\frac{1}{2}} dt = B\left(\frac{1}{2}+\frac{1}{n}, \frac{1}{2}\right)$$

$$= \frac{\Gamma\left(\frac{1}{n}+\frac{1}{2}\right)\Gamma\left(\frac{1}{2}\right)}{\Gamma\left(\frac{1}{n}+1\right)}$$

$$= \frac{\sqrt{\pi}\Gamma\left(\frac{1}{n}+\frac{1}{2}\right)}{\Gamma\left(\frac{1}{n}+1\right)}$$

Plugging this back into (14.8),

$$T = x_0^{1+\frac{n}{2}} \sqrt{\frac{m}{2nk}} \cdot \frac{\sqrt{\pi}\Gamma\left(\frac{1}{n}+\frac{1}{2}\right)}{\Gamma\left(\frac{1}{n}+1\right)}$$

$$= nx_0^{1+\frac{n}{2}} \sqrt{\frac{m\pi}{2nk}} \cdot \frac{\Gamma\left(\frac{1}{n}+\frac{1}{2}\right)}{\Gamma\left(\frac{1}{n}\right)}$$

Notice that as the point mass approaches the origin, v gets arbitrarily large (See (14.6)). However, we know that v can not become indefinitely large due to Einstein's theory of special relativity. Moreover, issues of relativistic mass have not been accounted for. So, this calculation is *not* relativistically correct. However, it is a good approximation and is a good exercise in classical mechanics.

Chapter 15

Physical Chemistry

15.1 Introduction

A **crystal structure** is an ordered arrangement of atoms, ions or molecules in a crystalline material[1]. The properties of any crystal structure mainly rely on the properties of the *unit cell*, which is the building block of any crystalline material. The unit cell is the smallest repeating unit having the symmetry of the full crystalline structure. The Coulombic forces between the constituents of the unit cell determine many properties[2].

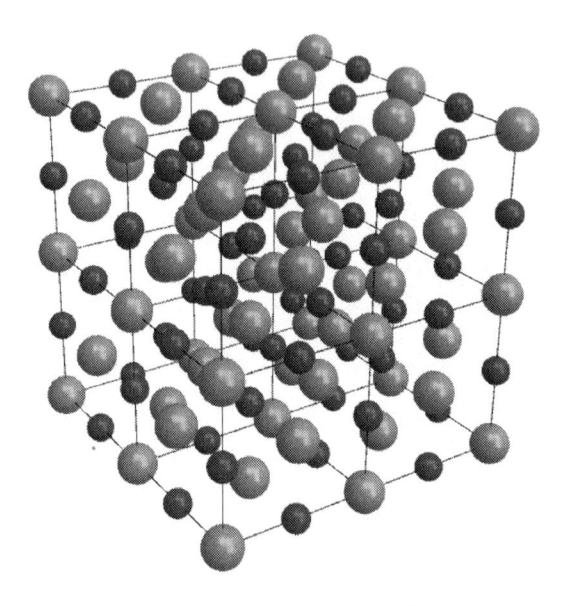

Figure 15.1: A visualization of the crystal structure of NaCl

The key quantity characterizing these Coulombic forces is called the **lattice energy**. The lattice energy of a crystalline solid is a measure of the energy released when ions are combined to

---
[1] Hook, J.R.; Hall, H.E. (2010). *Solid State Physics.* Manchester Physics Series (2nd ed.). John Wiley Sons. ISBN 9780471928041.
[2] S. Varughese, M. S. R. N. Kiran, U. Ramamurty and G. R. Desiraju. *Nanoindentation in Crystal Engineering: Quantifying Mechanical Properties of Molecular Crystals.* Angew. Chem. Int. Ed. 2013, 52, 2701-2712.

15.1. INTRODUCTION

make a compound. However, some sources such as the CRC Handbook of Chemistry and Physics define it oppositely, as the energy required to convert the crystal into infinitely separated gaseous ions in vacuum.

However, the lattice energy is not only determined by the interactions of the ions in the unit cell, but in the crystalline structure as a whole. The first efforts in calculating these long-range interactions were presented by the German physicist Erwin Madelung[3].

Madelung gave the electrostatic potential felt by a single ion in a crystal by approximating all ions as *point charges*. This quantity is named as *Madelung's constant*, and is used to determine the lattice energy through the **Born–Landé equation.**

> ### Scientific Model
>
> The Born-Landé equation is a means to calculate the lattice energy of a crystalline ionic compound. It was developed by the German physicist and mathematician Max Born as well as the German physicist Alfred Landé. It is given by
>
> $$E = \frac{NMz_+z_-e^2}{4\pi\epsilon_0 r}\left(\frac{1}{n} - 1\right) \quad (15.1)$$

Where

- $N = 6.02214076 \times 10^{23}$ denotes Avogadro's constant. As a refresher, this number denotes the number of constituents (molecules, atoms, formula units, etc.) that are present in one *mole*

- M is Madelung's constant

---
[3]Madelung E. (1918). *Das elektrische Feld in Systemen von regelmäßig angeordneten Punktladungen.* Phys. Z. XIX: 524-53

- z_+ is the charge number of the cation

- z_- is the charge number of the anion

- $e = 1.6022 \times 10^{19} C$ (Coulombs) is the elementary charge, i.e. the magnitude of the charge of a single electron.

- ϵ_0 is the vacuum permittivity

- r is the distance to the closest ion

- n is the *Born exponent*, which is experimentally determined

Partial Derivation. The electrostatic potential energy is

$$E = \frac{1}{4\pi\epsilon_0} \cdot \frac{q_1 q_2}{r}$$

For two charges q_1 and q_2 separated by a distance r. Note that when we say the charge of a sodium ion Na$^+$ is +1, we mean that it has a charge of e. Therefore,

$$E_0 = -\frac{1}{4\pi\epsilon_0} \cdot \frac{z_+ z_- e^2}{r}$$

For any compound. To account for the virtually infinite long-range interactions, we multiply by our multiplier M to get

$$E_{LR} = ME_0 = -\frac{M z_+ z_- e^2}{4\pi\epsilon_0 r}$$

Born and Landé suggested that the repulsive attractions between the lattice ions would be proportional to $\frac{1}{r^n}$ such that

15.1. INTRODUCTION

$$E_R = \frac{B}{r^n}$$

For a constant B. Thus,

$$E = E_{LR} + E_R = -\frac{Mz_+z_-e^2}{4\pi\epsilon_0 r} + \frac{B}{r^n} \qquad (15.2)$$

Where E denotes the total energy. Note that E must be at a minimum when $r = r_c$, or the equilibrium separation. This is due to the principle of energy minimization. To set that stipulation, we must find a constant B such that

$$\left.\frac{dE}{dr}\right|_{r=r_c} = 0$$

Therefore,

$$\frac{dE}{dr} = \frac{Mz_+z_-e^2}{4\pi\epsilon_0 r^2} - \frac{nB}{r^{n+1}}$$

$$\implies \frac{Mz_+z_-e^2}{4\pi\epsilon_0 r_c^2} - \frac{nB}{r_c^{n+1}} = 0$$

$$\frac{nB}{r_c^{n+1}} = \frac{Mz_+z_-e^2}{4\pi\epsilon_0 r_c^2}$$

$$\therefore B = \frac{Mz_+z_-e^2}{4\pi\epsilon_0 n} r_c^{n-1}$$

Plugging B in terms of r_c into (15.2) gives

$$E = -\frac{Mz_+z_-e^2}{4\pi\epsilon_0 r_c} + \frac{Mz_+z_-e^2}{4\pi\epsilon_0 n}r_c^{n-1}\cdot\frac{1}{r_c^n}$$

$$= \frac{Mz_+z_-e^2}{4\pi\epsilon_0 r_c}\left(\frac{1}{n}-1\right)$$

∎

Madelung's constant, although primarily of interest in inorganic chemistry, is also useful in describing the lattice energy of organic salts. Izgorodina et al. (2009) have described a generalized method, named the EUGEN method, for calculating the Madelung constant for any crystal structure[4].

In calculating the Madelung constant, we make a few assumptions about the ions in the crystal structure:

- The charges must have a spherically symmetric distribution of charge, i.e. a point charge or a charged metal sphere

- The charges must not overlap

- The charges must be stationary with respect to each other

Ions in a crystal lattice do not always have a spherically symmetric electron distribution, and therefore not a spherically symmetric distribution of charge.

Now that we know how the Madelung constant is used, we will explore it further. We can begin by defining it mathematically:

---

[4]E. Izgorodina; et al. (2009). *The Madelung Constant of Organic Salts*. Crystal Growth Design. 9 (11): 4834–4839. doi:10.1021/cg900656z.

15.1. INTRODUCTION

> **Definition**
>
> The Madelung constant of the nth ion, usually denoted as M_n, is defined as
>
> $$M_n = \sum_{n \neq k} \frac{z_k r_c}{r_{nk}} \qquad (15.3)$$
>
> Where r_{nk} denotes the distance between the nth and the kth ion, and the sum is taken over all ions in the lattice structure, which is virtually infinite.

Application. The electric potential felt by the nth ion is:

$$V_n = \frac{e}{4\pi\epsilon_0} \sum_{n \neq k} \frac{z_k}{r_{nk}}$$

Where z_k denotes the number charge of the kth ion. Notice that we sum over all k, so that we can account for the long-range interactions produced by every ion in the lattice. By normalizing r_{nk} to the distance to the closest ion, r_c, we obtain

$$V_n = \frac{e}{4\pi\epsilon_0 r_c} \sum_{n \neq k} \frac{z_k}{r_{nk}/r_c} = \frac{e}{4\pi\epsilon r_c} \sum_{n \neq k} \frac{z_k r_c}{r_{nk}}$$

By defining the sum above as Madelung's constant, we can write:

$$V_n = \frac{e}{4\pi\epsilon_0 r_c} M_n$$

It is here where we notice the utility of the Madelung constant. We can compute the electric potential energy produced by a virtually infinite array of ions by simply accounting for the electric potential generated by the closest ion. Notice that the electrostatic energy felt by the nth ion is the electric potential acting on the ion multiplied by the ion's charge,

$$E_n = V_n \cdot ez_n = \frac{e^2 z_n M_n}{4\pi\epsilon_0 r_c}$$

15.2 Sodium Chloride's Madelung Constant

Let us apply our new knowledge to the test! We will try to compute the Madelung constant for sodium chloride. We will first look at its structure.

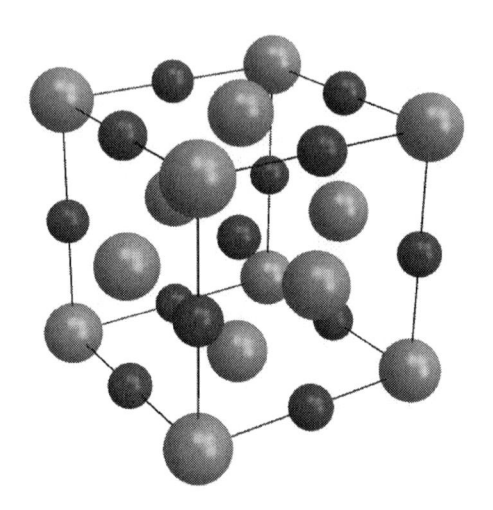

Figure 15.2: The unit cell of sodium chloride. Here, the blue atoms represent sodium while the bigger green atoms represent chlorine.

Notice that, because of sodium chloride's structure and symmetry, the Madelung constant has the same magnitude for both sodium and chlorine. However, the two different Madelung constants differ in sign, i.e.

15.2. SODIUM CHLORIDE'S MADELUNG CONSTANT 373

$$M_{\text{Na}} = -M_{\text{Cl}}$$

If we were to define a coordinate system, with the central atom in figure 17.1 being the origin, we can establish a series for the Madelung constant. We can also define the coordinate system such that half the *lattice constant* represents 1 unit. The lattice constant in this case is simply the side length of the cubic unit cell in figure 17.1. This is equivalent to what we did when we normalized the Madelung constant, as half the lattice constant is the distance to the closest ion.

We then need to sum over all possible integral (x, y, z) coordinates. By the Pythagorean theorem, the distance between the central atom at $(0,0,0)$ and some arbitrary ion located at (x, y, z) is

$$r = \sqrt{x^2 + y^2 + z^2}$$

Also, notice that the sodium and chlorine atoms alternate on every axis. To account for the difference in the sign of charge, we have to multiply by the factor $(-1)^{x+y+z}$. Since both sodium and chlorine have the same magnitude of charge (1), we need not worry about it. The Madelung constant is then

$$\sum_{(x,y,z)\neq(0,0,0)} \frac{(-1)^{x+y+z}}{\sqrt{x^2 + y^2 + z^2}} \tag{15.4}$$

But, how would we even go about evaluating this?

Neighbors	Distance
6 Cl$^-$	1
12 Na$^+$	$\sqrt{2}$
8 Cl$^-$	$\sqrt{3}$
6 Na$^+$	$\sqrt{4}$
24 Cl$^-$	$\sqrt{5}$

15.3 The Riemann Series Theorem in Action

Consider the central atom in figure 17.1. We will account for long range interactions by expanding spherically, i.e. we will consider ions a distance 1 away on our coordinate axis, then ions a distance $\sqrt{2}$ away, and so on. We can organize this information in a table. Therefore,

$$M_{Na} = 6 - \frac{12}{\sqrt{2}} + \frac{8}{\sqrt{3}} - \frac{6}{\sqrt{4}} + \frac{24}{\sqrt{5}} + \cdots \quad (15.5)$$

This is indeed the series given in many introductory textbooks[5,6]. However, this series diverges! This was proven by Emersleben in 1951[7]. We will outline his proof here:

Proposition. The series given in (15.5) diverges.

---

[5] J. S. Blakemore. *Solid State Physics.* Saunders, Philadelphia, 1969.
[6] C. Kittel. *Introduction To Solid State Physics.* Wiley, New York, 1953.
[7] Emersleben, O. (1951). *Das Selbstpotential einer endlichen Reihe neutraler äquidistanter Punktepaare.* Mathematische Nachrichten. 4 (3–4): 468. doi:10.1002/mana.3210040140.

15.3. THE RIEMANN SERIES THEOREM IN ACTION

Proof. Denote $r_3(n)$ as the number of representations of n as a sum of three squares. For example,

$$(\pm 1)^2 + 0^2 + 0^2 = 0^2 + (\pm 1)^2 + 0^2 = 0^2 + 0^2 + (\pm 1)^2 = 1$$

$$\implies r_3(1) = 3 \cdot 2 = 6$$

$r_3(n)$ is simply the denominator of the nth term in (15.5). Therefore,

$$M_{\text{Na}} = \sum_{n=1}^{\infty} \frac{(-1)^n r_3(n)}{\sqrt{n}} \qquad (15.6)$$

Now, denote L_R as the number of lattice points inside or on a sphere of radius R. This can be expressed as

$$L_R = \sum_{n=1}^{k} r_3(n) \qquad (15.7)$$

For $\sqrt{k} \leq R < \sqrt{k+1}$, $k \in \mathbb{N}^+$. We are essentially summing all lattice points by looking at layers of a sphere. It is easy to see that

$$L_R - \frac{4\pi R^3}{3} = O(R^2) \qquad (15.8)$$

For large R, the number of lattice points inside a sphere is approximately the volume of the sphere. Notice that we defined L_R as the number of lattice points inside *and* on the sphere. So the error term is only of magnitude r^2 because the surface area of a sphere is $4\pi r^2$. Since $\lim_{R \to \infty} \frac{R^2}{R^3} = 0$, we can write

$$\lim_{R \to \infty} \frac{L_R}{R^3} = \frac{4\pi}{3} \tag{15.9}$$

We will proceed to prove that (15.6) diverges by contradiction. Assume that (15.6) converges, then

$$\lim_{n \to \infty} \frac{r_3(n)}{\sqrt{n}} = 0$$

Now, consider letting $R = \sqrt{k}$ in (15.7) to get:

$$L_{\sqrt{k}} = \sum_{n=1}^{k} r_3(n)$$

Multiplying by $k^{-3/2}$,

$$\frac{L_{\sqrt{k}}}{k^{3/2}} = \frac{1}{k^{3/2}} \sum_{n=1}^{k} r_3(n)$$

Now, denote $X_n = \frac{r_3(n)}{\sqrt{n}}$. Thus,

$$\frac{L_{\sqrt{k}}}{k^{3/2}} = \frac{1}{k^{3/2}} \sum_{n=1}^{k} X_n \sqrt{n}$$

Let $k > N$ for some sufficiently large $N \in \mathbb{N}$. Also, let $C_N = \max\{X_n : n > N\}$, or the largest value of X_n on the interval $n \in (N, \infty)$. Splitting the sum,

$$\frac{L_{\sqrt{k}}}{k^{3/2}} = \frac{1}{k^{3/2}} \sum_{n=1}^{N} X_n \sqrt{n} + \frac{1}{k^{3/2}} \sum_{n=N+1}^{k} X_n \sqrt{n}$$

15.3. THE RIEMANN SERIES THEOREM IN ACTION

$$\leq \frac{1}{k^{3/2}} \sum_{n=1}^{N} X_n \sqrt{n} + \frac{C_N}{k^{3/2}} \sum_{n=N+1}^{k} \sqrt{n} \qquad (15.10)$$

By the integral test (See Chapter 5),

$$\sum_{n=N+1}^{k} \sqrt{n} \leq \int_{N+1}^{k+1} \sqrt{x} \, dx = \left[\frac{2}{3} x^{3/2}\right]_{N+1}^{k+1}$$

$$= \frac{2}{3}\left((k+1)^{3/2} - (N+1)^{3/2}\right) \qquad (15.11)$$

Substituting (15.11) into (15.10) gives:

$$\frac{L_{\sqrt{k}}}{k^{3/2}} \leq \frac{1}{k^{3/2}} \sum_{n=1}^{N} X_n \sqrt{n} + \frac{2C_N}{3k^{3/2}} \left((k+1)^{3/2} - (N+1)^{3/2}\right)$$

$$= \frac{1}{k^{3/2}} \sum_{n=1}^{N} X_n \sqrt{n} + \frac{2C_N}{3} \left(\left(\frac{k+1}{k}\right)^{3/2} - \left(\frac{N+1}{k}\right)^{3/2}\right)$$

Letting $k \to \infty$ gives:

$$\limsup_{k \to \infty} \frac{L_{\sqrt{k}}}{k^{3/2}} \leq \frac{2C_N}{3}$$

By our assumption that (15.5) converges and the definition of convergence, we necessarily have

$$\lim_{N \to \infty} C_N = 0$$

Thus,

$$\lim_{k \to \infty} \frac{L_{\sqrt{k}}}{k^{3/2}} = 0$$

However, this is a contradiction of (15.9)! Therefore, (15.5) diverges.

□

How could this happen? Physically, we observe these long-range interactions as defined and finite. Could there be a flaw with our physics in deriving the Madelung constant? Well, no. The series given in (15.4) is only *conditionally convergent*. Therefore, by the Riemann series theorem, the order in which we arrange the series impacts its value. As a matter of fact, a conditionally convergent series can be made to diverge, or fail to approach any limit!

The definition given in (15.3) does not specify how to go about summing the virtually infinite terms. Therefore, we need to go about summing series in a way that is physically meaningful as well as physically accurate.

Perhaps, the method we used to sum our series (expanding spheres) is incorrect. In retrospect, the expanding spheres method has no physical meaningfulness since there are no spherical crystals. Nonetheless, its simplicity makes it attractive. The correct way to sum the series in (15.4) is by the method of **expanding cubes**. An elementary proof for the convergence of this method is given by Borwein et Al (1985)[8].

Unfortunately, there is no closed form for (15.4)[9]. The method

---

[8]Borwein, D.; Borwein, J. M.; Taylor, K. F. (1985). *Convergence of Lattice Sums and Madelung's Constant.* J. Math. Phys. 26 (11): 2999–3009. Bibcode:1985JMP....26.2999B. doi:10.1063/1.526675.

[9]Bailey, D. H.; Borwein, J. M.; Kapoor, V.; and Weisstein, E. W. *Ten Problems in Experimental Mathematics.* Amer. Math. Monthly 113, 481-509, 2006.

of expanding cubes gives the value:

$$\sum_{(x,y,z)\neq(0,0,0)} \frac{(-1)^{x+y+z}}{\sqrt{x^2+y^2+z^2}} = 1.747\ldots$$

Which is experimentally verified. However, the rate of convergence of this sum using straightforward methods is extremely slow. Thousands of steps are required to obtain accuracy to one decimal place.

In order to accelerate computation, Bailey et Al (2006) gave the following rapidly converging series

$$\sum_{(x,y,z)\neq(0,0,0)} \frac{(-1)^{x+y+z}}{\sqrt{x^2+y^2+z^2}} = 12\pi \sum_{n,k\geq 1,\,\text{odd}} \operatorname{sech}^2\left(\frac{\pi}{2}(n^2+k^2)^{1/2}\right)$$

Unfortunately, the methods employed in their derivation, such as Jacobi theta functions and Mellin transforms, are outside the scope of this book.

15.4 Pharmaceutical Connections

Even though most of the properties of a crystal structure can be accurately assessed by looking at the short-range structure and interactions, it is important to note that long range Coulombic attractions are important and necessary for accuracy. This is especially relevant in the pharmaceutical industry, where long range interactions affect the "bulk behavior" of crystals. Ho et Al. (2019) state that "API (Active Pharmaceutical Ingredient) physical/chemical properties across scales can significantly influence formulation choice/composition, manufacturing route, and performance of the pharmaceutical product."[10] This keys us in to the importance of the Madelung constant, and how the mathematics here indeed translates into various fields of science!

15.5 The Debye Model

> **Definition**
>
> The Debye model is a model developed by Dutch-American physicist and chemist Peter Debye to predict the specific heat of solids[a]. This model correctly predicts the temperature dependence of the heat capacity for low temperatures and high temperatures, but fails at intermediate temperatures due to its underlying assumptions.
>
> [a]Debye, Peter (1912). *Zur Theorie der spezifischen Waerme*. Annalen der Physik. 39 (4): 789–839. Bibcode:1912AnP...344..789D. doi:10.1002/andp.19123441404.

The dimensionless heat capacity is given by

$$C = 9Nk \left(\frac{T}{T_D}\right)^3 \int_0^{T_D/T} \frac{x^4 e^x}{(e^x - 1)^2} dx$$

Where

- N is the number of atoms
- k is Boltzmann's constant

[10]Ho, Raimundo, et al. *Multiscale Assessment Of Api Physical Properties In The Context Of Materials Science Tetrahedron Concept*. Chemical Engineering in the Pharmaceutical Industry, 2019, pp. 689–712., doi:10.1002/9781119600800.ch30.

15.5. THE DEBYE MODEL

- T_D is the Debye temperature, which is the highest temperature in a lattice that can be achieved due to a single normal vibration

- T is the temperature

At low temperatures,
$$\frac{T_D}{T} \to \infty$$
Hence,

$$C = 9Nk \left(\frac{T}{T_D}\right)^3 \int_0^\infty \frac{x^4 e^x}{(e^x - 1)^2} dx$$

This integral can be easily evaluated. We start by using IBP with $u = x^4$, $dv = \frac{e^x}{(e^x-1)^2} dx$:

$$\int_0^\infty \frac{x^4 e^x}{(e^x - 1)^2} dx = \left[-\frac{x^4}{e^x - 1}\right]_0^\infty - \int_0^\infty \frac{4x^3}{e^x - 1} dx$$

We first focus on the evaluation of the first term. Its evaluation at $x = \infty$ can be trivially shown to be equal to 0, but its evaluation at $x = 0$ requires a little more work. Using L'Hopital's rule gives us

$$\lim_{x \to 0} -\frac{x^4}{e^x - 1} = \lim_{x \to 0} -\frac{\frac{d}{dx} x^4}{\frac{d}{dx}(e^x - 1)}$$

$$= -\lim_{x \to 0} 4e^{-x} x^3$$

Simply plugging in $x = 0$ then gives us that the first term is $0 + 0 = 0$. Plugging this back into our expression for C,

$$C = 36Nk \left(\frac{T}{T_D}\right)^3 \int_0^\infty \frac{x^3}{e^x - 1} dx$$

Although this form was already given in (12.5), we can consider another approach. We can multiply the integral by e^{-x} to get,

$$I = \int_0^\infty \frac{x^3 e^{-x}}{1 - e^{-x}} dx$$

Using the power series

$$\frac{1}{1-x} = \sum_{n=0}^\infty x^n$$

We can express I as:

$$I = \int_0^\infty x^3 e^{-x} \sum_{n=0}^\infty e^{-xn} dx$$

Interchanging summation and integration by the dominated convergence theorem,

$$I = \sum_{n=1}^\infty \int_0^\infty x^3 e^{-xn} dx$$

Integrating by parts then gives

$$I = 3! \sum_{n=1}^\infty \frac{1}{n^4}$$

$$= 6\zeta(4) = \frac{\pi^4}{15}$$

15.5. THE DEBYE MODEL

Which agrees with our result if we had used (12.5). Plugging this integral into the expression for C then gives

$$C = \frac{12\pi^4 Nk}{5} \left(\frac{T}{T_D}\right)^3$$

Chapter 16

Statistical Mechanics

16.1 Introduction

Statistical mechanics may be a new field for the reader. Nonetheless, it is one of the pillars of modern physics, along with general relativity and quantum mechanics. Primarily, the concepts of the field are necessary for any calculations in physical systems that have a large number of **degrees of freedom**.

> **Definition**
>
> In statistical mechanics, a degree of freedom is a scalar quantity describing the **microstate** of the system. A microstate is simply a specific microscopic configuration of a thermodynamic system.

The field derives from concepts in statistics as well as physical principles[1]. To describe a system. statistical mechanics introduces the notion of a **statistical ensemble** which was introduced by the American scientist Josiah Willard Gibbs in 1902[2]. In Gibbs' seminal book *Elementary Principles in Statistical Mechanics*, he accounted for several thermodynamic properties of large systems in terms of the statistics of ensembles. This, along with the works of James Clerk Maxwell and Ludwig Boltzmann, established the foundations of statistical mechanics.

> **Definition**
>
> The statistical ensemble is a collection of many possible states of a system, each assigned a certain probability. Simply put, a statistical ensemble is a probability distribution for the state of the system.

---

[1] Tolman, R. C. (1938). *The Principles of Statistical Mechanics*. Dover Publications. ISBN 9780486638966.
[2] Gibbs, Josiah Willard (1902). *Elementary Principles in Statistical Mechanics*. New York: Charles Scribner's Sons.

16.2. EQUATIONS OF STATE

Now, there are three main *ensembles* defined for any isolated system occupying a finite volume:

- Microcanonical ensemble
- Canonical ensemble
- Grand canonical ensemble

These ensembles are the primary means by which statistical mechanics derives its theoretical predictions. After a brief introduction to this field, we will now transition into the applications of the mathematics we have learned!

16.2 Equations of State

The reader might recall the ideal gas law from chemistry,

$$PV = nRT \tag{16.1}$$

Where:

- P denotes the pressure
- V denotes the volume
- n denotes the number of moles of gas
- R denotes the ideal gas constant
- T denotes the absolute temperature (in kelvins)

The equation was first stated by the French physicist Émile Clapeyron as a combination of the less inclusive gas laws of

388   CHAPTER 16. STATISTICAL MECHANICS

Boyle, Charles, Avogadro, and Gay-Lussac[3]. This equation is **the equation of state** of the hypothetical ideal gas. An ideal gas is a theoretical gas composed of a large number of randomly moving point particles whose only interactions are perfectly elastic collisions. However, this is far from true, especially for gases in high pressure, low temperature settings. Furthermore, gases with large particles also deviate from ideal behavior.

Nonetheless, the ideal gas serves as a first model in thermodynamics and statistical mechanics. In more advanced curricula, corrections to this model are introduced. One correction is the compressibility factor.

> **Definition**
>
> The compressibly factor, usually denoted as Z, is a correction factor which describes the deviation of a real gas from ideal gas behaviour. It is defined as:
>
> $$Z = \frac{PV}{nRT}$$
>
> Notice that for an ideal gas, $Z = 1$ by the ideal gas formula (16.1).

Before we proceed, we need to make note of statistical mechanics' use of the *number of particles* instead of the number of moles. Moreover, note that the ideal gas constant is given by:

$$R = Ak$$

Where A denotes Avogadro's constant and k denotes Boltzmann's constant. Therefore, we can express the ideal gas law as

---

[3]Clapeyron, E. (1834). *Mémoire sur la puissance motrice de la chaleur.* Journal de l'École Polytechnique (in French). XIV: 153–90. Facsimile at the Bibliothèque nationale de France (pp. 153–90).

$$PV = NkT$$

Where N denotes the number of particles.

16.3 Virial Expansion

One way to approximate Z is by using the virial expansion. We will begin by defining the particle density as

$$\rho \equiv \frac{N}{V}$$

$$\implies \frac{P}{kT} = \rho$$

Now, consider writing Z as a power series in ρ:

$$Z = 1 + B_2\rho + B_3\rho^2 + \cdots$$

The coefficients $B_2, B_3, \cdots$ are often represented by Taylor series in $\frac{1}{T}$. The biggest correction to ideal gas behavior will be our first coefficient, B_2. It is given by

$$B_2(T) = -2\pi \int r^2 \left(e^{-u(r)/(kT)} - 1\right) dr$$

Where $u(r)$ gives the potential energy at r.

16.3.1 Lennard-Jones Potential

> **Scientific Model**
>
> The Lennard-Jones potential is a simple model that approximates the interaction between a pair of neutral atoms or molecules. A form of this interatomic potential was first proposed in 1924 by John Lennard-Jones[a]. The most common version of this model is given by:
>
> $$V_{\text{LJ}}(r) = 4\epsilon \left[\left(\frac{\sigma}{r}\right)^{12} - \left(\frac{\sigma}{r}\right)^{6} \right]$$
>
> ---
>
> [a]Lennard-Jones, J. E. (1924). *On the Determination of Molecular Fields*. Proc. R. Soc. Lond. A, 106 (738): 463–477. Bibcode:1924RSPSA.106..463J, doi:10.1098/rspa.1924.0082

Where

- ϵ is the depth of the potential well
- σ is the finite distance at which the inter-particle potential is zero
- r is the distance between the particles

Due to its mathematical simplicity, the Lennard-Jones potential is used extensively in computer simulations of chemical phenomena even though more accurate models exist. The r^{-12} term, which is the repulsive term, describes Pauli repulsion at short ranges due to overlapping electron orbitals (See Pauli exclusion principle), and the r^{-6} term describes attraction at long ranges due to electric dipole fluctuations (Van der Waals force, or dispersion force).

With this interaction potential, the first virial coefficient is given by:

16.3. VIRIAL EXPANSION

$$B_2 = -2\pi \int_0^\infty r^2 \left[\exp\left(-\frac{V_{\text{LJ}}(r)}{kT}\right) - 1\right] dr$$

$$= -2\pi \int_0^\infty r^2 \left[\exp\left(-\frac{4\epsilon}{kT}\left[\left(\frac{\sigma}{r}\right)^{12} - \left(\frac{\sigma}{r}\right)^6\right]\right) - 1\right] dr$$

Where exp denotes the exponential function, i.e. $\exp(x) = e^x$.
Substituting $x = \frac{r}{\sigma}$, $dx = \frac{1}{\sigma}dr$ gives:

$$B_2 = -2\pi\sigma^3 \int_0^\infty x^2 \left[\exp\left(-\frac{4\epsilon}{kT}\left[\frac{1}{x^{12}} - \frac{1}{x^6}\right]\right) - 1\right] dx$$

Substituting again with $T^* = \frac{kT}{\epsilon}$,

$$B_2 = -2\pi\sigma^3 \int_0^\infty x^2 \left[\exp\left(-\frac{4}{T^*}\left[\frac{1}{x^{12}} - \frac{1}{x^6}\right]\right) - 1\right] dx$$

Challenge Problem

Show that the above expression can be expressed as[a]:

$$B_2 = -2\sum_{n=1}^\infty \frac{1}{4n!}\Gamma\left(\frac{2n-1}{4}\right)\left(\frac{1}{T^*}\right)^{(2n+1)/4}$$

This series converges rapidly for $T^* > 4$.

---
[a] Reichl, L. E. (2017). *A modern course in statistical physics*. Weinheim: Wiley-VCH.

16.4 Blackbody Radiation

Every object spontaneously and continuously emits electromagnetic radiation. To be more precise, any object with a temperature *above* 0 kelvin emits electromagnetic radiation. This radiation is a distribution of light with different frequencies. The hotter the object, the higher frequencies it generates. That is why when a piece of metal is heated, it starts to glow with a mildly dull red color and then progresses to a brighter yellow, a color with a higher frequency.

Even the person reading this right now is generating light! However, much of this light is in the infrared spectrum, making it invisible to the unassisted human eye. Nonetheless, it is still energy that can be harnessed for power.

Conversely, all objects absorb electromagnetic radiation to some degree as well. We define an object that absorbs all radiation falling on it, at all wavelengths, as a **black body**.

The spectral radiance, B_ν, of a body describes the amount of energy it emits at different radiation frequencies. When a black body is at a uniform temperature, its emission has a characteristic frequency distribution that depends on the temperature. This emission is called black-body radiation.

Now, we go back about 100 years to the year 1900. The model used to calculate the spectral radiance of a black body was the **Rayleigh-Jeans law** given by

$$B_\lambda(T) = \frac{2ckT}{\lambda^4}$$

Where

- λ is the wavelength

16.4. BLACKBODY RADIATION

- c is the speed of light
- k is Boltzmann's constant
- T is the absolute temperature

We can also write the Rayleigh-Jeans law in terms of the frequency, ν,

$$B_\nu(T) = \frac{2\nu^2 kT}{c^2}$$

The British physicist Lord Rayleigh derived the λ^{-4} dependence of the spectral radiance based on both classical physical arguments and empirical facts[4]. Then, a more complete derivation that included the proportionality constant was presented by Rayleigh and Sir James Jeans in 1905.

Notice the word *classical* in the above paragraph. As it turns out, this theory fails and demonstrates a key error in the understanding of physics at the time. As for why the theory is erroneous, recall that a black body emits a *distribution* of light frequencies. So, we might perhaps want to integrate $B_\nu(T)$ over all ν to gauge how much energy an object is emitting

$$\int_0^\infty B_\nu(T)\,\mathrm{d}\nu = \int_0^\infty \frac{2\nu^2 kT}{c^2}\mathrm{d}\nu$$

$$\frac{2kT}{c^2}\left[\frac{\nu^3}{3}\right]_0^\infty = \infty$$

Which diverges! But, this can not be true. The energy an object is emitting must be finite. This theoretical malfunction

---
[4]Kutner, Mark L. (2003). *Astronomy: A Physical Perspective*. Cambridge University Press. p. 15. ISBN 0-521-52927-1.

would be named the **ultraviolet catastrophe**. The problem is, the law correctly predicts experimental results for $\nu < 10^5$ GHz but fails to do so as frequencies reach the ultraviolet region of the electromagnetic spectrum.

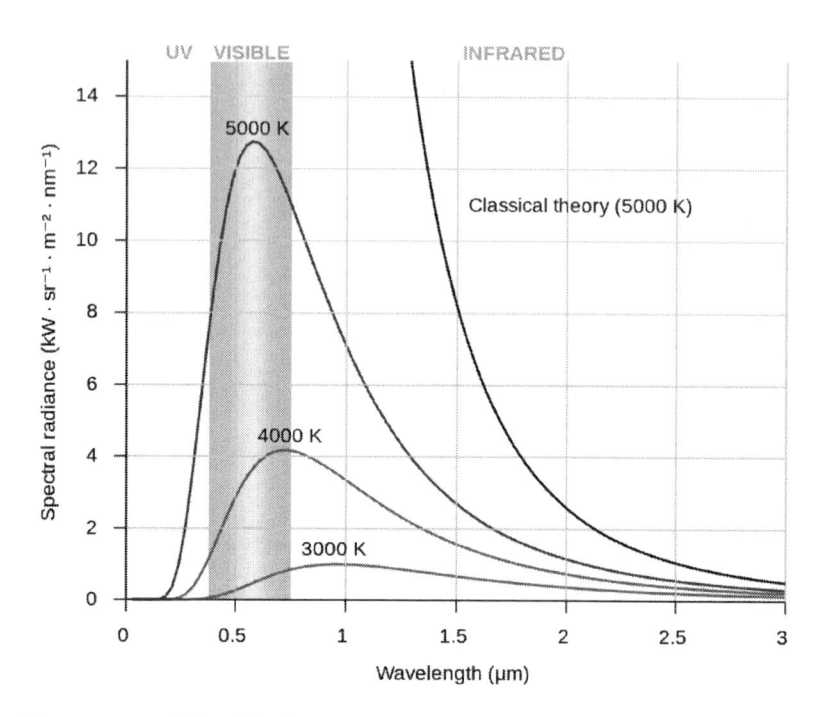

Figure 16.1: The black curve, representing the Rayleigh-Jeans law, diverges from the observed intensity. Image by Darth Kule - Own work, Public Domain, https://commons.wikimedia.org/w/index.php?curid=10555337

Enter stage right the revolutionary German physicist Max Planck, also known as the father of quantum mechanics (Which was due to this topic!). Planck showed that the spectral radiance of a body for frequency ν at absolute temperature T is given by

$$B_\nu(\nu, T) = \frac{2h\nu^3}{c^2} \frac{1}{e^{\frac{h\nu}{kT}} - 1}$$

16.4. BLACKBODY RADIATION

Where h denotes the Planck constant. Note that B_ν has the unit "power per unit surface area per unit solid angle per unit frequency emitted." To derive the power emitted per unit area, we have to calculate the integral

$$\frac{P}{A} = \int_0^\infty B_\nu d\nu \int_0^{2\pi} d\phi \int_0^{\pi/2} \cos\theta \sin\theta \, d\theta$$
$$= \pi \int_0^\infty B_\nu d\nu$$

Substituting $x = \frac{h\nu}{kT}$, $d\nu = \frac{kT}{h} dx$ gives:

$$\int_0^\infty B_\nu \, d\nu = \left(\frac{2h}{c^2}\right) \left(\frac{kT}{h}\right)^4 \int_0^\infty \frac{x^3}{e^x - 1} dx$$

It is here where we see that we can use (16.2). However, we can derive that formula here! We begin by letting

$$f(\alpha) = \int_0^\infty \frac{x^\alpha}{e^x - 1} dx$$

Now, consider the integral definition of the Gamma function

$$\Gamma(z) = \int_0^\infty x^{z-1} e^{-x} dx$$

Then the substitution $x = nu$, $dx = n \, du$ for some $n \in \mathbb{N}^+$ yields:

$$\Gamma(z) = \int_0^\infty n^z u^{z-1} e^{-nu} du$$

$$\therefore \frac{1}{n^z} = \frac{1}{\Gamma(z)} \int_0^\infty u^{z-1} e^{-nu} du$$

We can now sum both sides from $n = 1$ to ∞.

$$\sum_{n=1}^{\infty} \frac{1}{n^z} = \sum_{n=1}^{\infty} \frac{1}{\Gamma(z)} \int_0^{\infty} u^{z-1} e^{-nu} du$$

By the dominated convergence theorem, we can interchange summation and integration to get

$$\sum_{n=1}^{\infty} \frac{1}{n^z} = \frac{1}{\Gamma(z)} \int_0^{\infty} u^{z-1} \sum_{n=1}^{\infty} e^{-nu} du$$

We can then easily evaluate the sum on the RHS by the power series expansion of $\frac{1}{1-x}$.

$$\implies \sum_{n=1}^{\infty} \frac{1}{n^z} = \frac{1}{\Gamma(z)} \int_0^{\infty} u^{z-1} \left(\frac{1}{1-e^{-u}} - 1 \right) du$$

$$= \frac{1}{\Gamma(z)} \int_0^{\infty} \frac{u^{z-1} e^{-u}}{1-e^{-u}} du$$

We can multiply by $\frac{e^u}{e^u}$ to get:

$$\sum_{n=1}^{\infty} \frac{1}{n^z} = \frac{1}{\Gamma(z)} \int_0^{\infty} \frac{u^{z-1}}{e^u - 1} du$$

Notice that for $\Re(z) > 1$, the RHS is equivalent to $\zeta(z)$. Therefore,

$$\zeta(z)\Gamma(z) = \int_0^{\infty} \frac{u^{z-1}}{e^u - 1} du \qquad (16.2)$$

Which is a beautiful formula. It gives rise to this interesting result,

$$\int_0^\infty \frac{u}{e^u - 1} du = \frac{\pi^2}{6}$$

We can then use (16.2) to determine that

$$\int_0^\infty \frac{x^3}{e^x - 1} dx = \zeta(4)\Gamma(4) = 6 \cdot \frac{\pi^4}{90} = \frac{\pi^4}{15}$$

Back to our original problem, we have that

$$\frac{P}{A} = \frac{\pi^5}{15} \left(\frac{2h}{c^2}\right) \left(\frac{kT}{h}\right)^4$$

$$= \left(\frac{2\pi^5 k^4}{15 h^3 c^2}\right) T^4$$

This is the **Stefan–Boltzmann law**, which is used in a variety of applications. Markedly, it is used in astrophysics to calculate the temperature of stars!

16.5 Fermi-Dirac (F-D) Statistics

Fermi–Dirac statistics describe a distribution of particles over energy states in systems consisting of many identical particles that obey the "Pauli exclusion principle". It is named after the Italian physicist Enrico Fermi and English physicist Paul Dirac[5,6].

---

[5] Fermi, Enrico (1926). *Sulla quantizzazione del gas perfetto monoatomico*. Rendiconti Lincei (in Italian). 3: 145–9., translated as Zannoni, Alberto (1999-12-14). *On the Quantization of the Monoatomic Ideal Gas*. arXiv:cond-mat/9912229.

[6] Dirac, Paul A. M. (1926). *On the Theory of Quantum Mechanics*. Proceedings of the Royal Society A. 112 (762): 661–77. Bibcode:1926RSPSA.112..661D. doi:10.1098/rspa.1926.0133. JSTOR 94692.

Scientific Model

The Fermi-Dirac (F-D) distribution characterizes a system of identical **fermions** (particles with half integer spin) in thermodynamic equilibrium. In that scenario, the average number of fermions in a single-particle state i is given by the sigmoid function[a]:

$$n_i = \frac{1}{e^{(E_i - \mu)/kT} + 1} \qquad (16.3)$$

[a] Reif, F. (1965). *Fundamentals of Statistical and Thermal Physics*. McGraw–Hill. ISBN 978-0-07-051800-1.

Where

- k is Boltzmann's constant

- T is the absolute temperature

- E_i is the energy of a single-particle state i

- μ is the total chemical potential

Fermions include all quarks and leptons, as well as all composite particles made of an odd number of these. They can be either elementary particles (electrons) or composite particles (protons). The F-D distribution is most frequently applied to electrons, which have spin $\frac{1}{2}$. The F–D distribution is *only valid* if the number of fermions in the system is large enough so that adding fermions has a negligible effect on μ.

Derivation. We begin by noting the Pauli Exclusion Principle:

16.5. FERMI-DIRAC (F-D) STATISTICS

> **Scientific Law**
>
> The Pauli exclusion principle was formulated by Austrian physicist Wolfgang Pauli in 1925 for electrons, and later extended to all fermions with his spin–statistics theorem of 1940[a]. It states that two or more identical fermions cannot occupy the same quantum state within a quantum system simultaneously.
>
> ---
> [a]Pauli, Wolfgang (1980). *General principles of quantum mechanics*. Springer-Verlag. ISBN 9783540098423.

So we have the restriction that in our system, each allowed energy state s_i can accommodate one and only one fermion. We can also set the restrictions that

- N, the number of fermions in the system, is a fixed number
- $\sum_i E_i N_i$, the total energy of the system, is constant

Now, consider the problem of placing N_i indistinguishable fermions in the ith level into S_i states. This is simply:

$$W_i = \binom{S_i}{N_i} = \frac{S_i!}{(S_i - N_i)!\, N_i!}$$

Therefore, the total number of ways to arrange indistinguishable fermions in a multi-level system is

$$W = \prod_i W_i = \prod_i \frac{S_i!}{(S_i - N_i)!\, N_i!}$$

Where the product is taken over all i. We now seek to set N_i values such that we maximize W using our basic optimization tools. Remember from the fundamental laws of thermodynamics that the most *likely* distribution is that with the largest amount of microstates. But, before we proceed we need to take the natural logarithm of W. Simply put, solving $\mathrm{d}\left(\ln W\right) = 0$ is much simpler than solving $\mathrm{d}(W) = 0$. Since

$$\mathrm{d}\left(\ln W\right) = \frac{\mathrm{d}(W)}{W}$$

And $W \neq 0$, we are justified in this step. Taking the natural logarithm of W,

$$\ln W = \sum_i \ln\left(S_i!\right) - \ln\left((S_i - N_i)!\right) - \ln\left(N_i!\right)$$

Since we are dealing with a large number of fermions, we can use Stirling's formula from (1.5).

$$\ln x! = x \ln x - x$$

For large x. Applying this approximation gives:

$$\ln W = \sum_i S_i \ln\left(S_i\right) - S_i - \left(S_i - N_i\right)\ln\left(S_i - N_i\right)$$

$$+ \left(S_i - N_i\right) - N_i \ln\left(N_i\right) + N_i$$

$$= \sum_i S_i \ln\left(S_i\right) - \left(S_i - N_i\right)\ln\left(S_i - N_i\right) - N_i \ln\left(N_i\right)$$

Now, we are in a comfortable place to perform our optimization.

16.5. FERMI-DIRAC (F-D) STATISTICS

$$\frac{d\ln W}{dN_i} = \sum_i \frac{\partial W}{\partial N_i}$$

$$d(\ln W) = \sum_i \frac{\partial W}{\partial N_i} dN_i \qquad (16.4)$$

Since $\frac{dS_i}{dN_i} = 0$ (S_i is a constant with respect to N_i), we can write:

$$\frac{\partial W}{\partial N_i} = \ln(S_i - N_i) + 1 - \ln(N_i) - 1$$

$$= \ln\left(\frac{S_i}{N_i} - 1\right) \qquad (16.5)$$

Plugging this value into (16.4),

$$d(\ln W) = \sum_i \left[\ln\left(\frac{S_i}{N_i} - 1\right)\right] dN_i$$

Now, our optimization step.

$$\sum_i \left[\ln\left(\frac{S_i}{N_i} - 1\right)\right] dN_i = 0$$

But, we have the restrictions:

- There is a fixed number of particles
- The energy of the system is constant

Although many readers are familiar with how to deal with such constrained problems (Using Lagrange multipliers), a brief introduction to Lagrange multipliers will be provided.

Theorem

The method of Lagrange multipliers is a strategy for finding the local maxima or minima of a function subject to equality constraints[a]. The method for one constraint can be summarized as follows: in order to find the stationary points of $f(x)$ under the constraint $g(x) = 0$, find the stationary points of:

$$\mathcal{L}(x, \lambda) = f(x) + \lambda g(x)$$

Where λ is the **Langrage multiplier**.

---
[a] Hoffmann, Laurence D.; Bradley, Gerald L. (2004). *Calculus for Business, Economics, and the Social and Life Sciences (8th ed.)*. pp. 575–588. ISBN 0-07-242432-X.

This method is advantageous since it allows optimization problems to be solved without explicit parameterization in terms of the constraints. Therefore, the method of Lagrange multipliers is very popular in solving challenging constrained optimization problems. Using this method gives the two constraints:

$$N - \sum_i N_i = 0$$

$$E - \sum_i E_i N_i = 0$$

Therefore,

$$\mathcal{L}(x, \lambda_1, \lambda_2) = \ln W + \lambda_1 \left(N - \sum_i N_i \right) + \lambda_2 \left(E - \sum_i E_i N_i \right)$$

In general, we can maximize W using $\frac{d\mathcal{L}}{dN_i} = 0$,

16.5. FERMI-DIRAC (F-D) STATISTICS

$$\frac{\partial W}{\partial N_i} = \lambda_1 + \lambda_2 E_i$$

Using (16.5) and exponentiating both sides,,

$$\ln\left(\frac{S_i}{N_i} - 1\right) = \lambda_1 + \lambda_2 E_i$$

$$\frac{S_i}{N_i} - 1 = e^{\lambda_1 + \lambda_2 E_i}$$

$$\therefore N_i = \frac{S_i}{1 + e^{\lambda_1 + \lambda_2 E_i}}$$

For some constants λ_1, λ_2. Now, our focus shifts to trying to evaluate λ_1, λ_2. Recall Boltzmann's law, which states that:

$$S = k \ln W = k\mathcal{L}(x, \lambda_1, \lambda_2)$$

Where S denotes entropy. We can then see that

$$\frac{\partial S}{\partial N} = k\lambda_1 \ , \ \frac{\partial S}{\partial E} = k\lambda_2$$

We can now utilize the thermodynamic equation

$$dE = TdS - PdV + \mu dN$$

$$\implies dS = \frac{1}{T}(dE + PdV - \mu dN)$$

$$\implies \left(\frac{\partial S}{\partial N}\right)_{E,V} = k\lambda_1 = -\frac{\mu}{T}$$

$$\left(\frac{\partial S}{\partial E}\right)_{N,V} = k\lambda_2 = \frac{1}{T}$$

Solving for λ_1, λ_2

$$\lambda_1 = -\frac{\mu}{kT}$$

$$\lambda_2 = \frac{1}{kT}$$

Therefore,

$$n_i = \frac{1}{e^{(E_i-\mu)/kT} + 1}$$

Chapter 17

Miscellaneous

17.1 Volume of a Hypersphere of Dimension N

The N deminsional hypersphere, or N−ball, is a generalization of the notion of a sphere to higher dimensions. A a circle of radius R (which can be regarded as a 2−ball) has the equation:

$$x^2 + y^2 \leq R^2$$

Similarly, for 3 dimensions, the equation of a sphere of radius R is given by:

$$x^2 + y^2 + z^2 \leq R^2$$

In general, for an N−ball, its volume is the volume enclosed by

$$x_1^2 + x_2^2 + \cdots + x_N^2 \leq R^2$$

17.1.1 Spherical Coordinates

To the reader that is familiar with multiple integrals and the use of spherical coordinates, this section will be rather easy. However, a brief introduction is given below.

17.1. VOLUME OF A HYPERSPHERE OF DIMENSION N

> **Definition**
>
> The spherical coordinates of a point are given by (r, θ, φ), where
>
> - $r = \sqrt{x^2 + y^2 + z^2}$
> - $\varphi = \arctan\left(\frac{y}{x}\right)$
> - $\theta = \arccos\left(\frac{z}{r}\right)$
>
> And r, θ, φ denote radius, inclination, and azimuth, respectively.

Conversely, one can write:

- $x = r \sin\theta \cos\varphi$
- $y = r \sin\theta \sin\varphi$
- $z = r \cos\theta$

We can extend this spherical coordinate system to an arbitrary dimension. By using the notion of the Jacobian,

$$d^n V = \left| \frac{\partial(x_1, x_2, \cdots, x_n)}{\partial(r, \varphi_1, \cdots, \varphi_{n-1})} \right|$$

$$= \det \begin{pmatrix} \cos(\varphi_1) & -r\sin(\varphi_1) & 0 & 0 & \cdots & 0 \\ \sin(\varphi_1)\cos(\varphi_2) & r\cos(\varphi_1)\cos(\varphi_2) & -r\sin(\varphi_1)\sin(\varphi_2) & 0 & \cdots & 0 \\ \vdots & \vdots & \vdots & \cdots & & \vdots \\ & & & & & 0 \\ \sin(\varphi_1)\cdots\sin(\varphi_{n-2})\cos(\varphi_{n-1}) & \cdots & \cdots & & -r\sin(\varphi_1)\cdots\sin(\varphi_{n-2})\sin(\varphi_{n-1}) \\ \sin(\varphi_1)\cdots\sin(\varphi_{n-2})\sin(\varphi_{n-1}) & r\cos(\varphi_1)\cdots\sin(\varphi_{n-1}) & \cdots & & r\sin(\varphi_1)\cdots\sin(\varphi_{n-2})\cos(\varphi_{n-1}) \end{pmatrix}$$

$$= r^{n-1} \sin^{n-2}(\varphi_1) \sin^{n-3}(\varphi_2) \cdots \sin(\varphi_{n-2}) \, dr \, d\varphi_1 \, d\varphi_2 \cdots d\varphi_{n-1}$$

Where $|\cdot|$ denotes the determinant. In this coordinate system, there is one radial coordinate, r, and $n-1$ angular coordinates.

The domain of each angular coordinate is $[0, \pi]$, except ϕ_{n-1} which has domain $[0, 2\pi]$. A more detailed derivation can be found in the Blumenson (1960)[1].

17.1.2 Calculation

Using spherical coordinates, the volume of an N-ball is

$$V_N(R) = \int_0^R \int_0^\pi \int_0^\pi \cdots \int_0^{2\pi} r^{N-1} \sin^{N-2}(\varphi_1) \cdots$$
$$\sin(\varphi_{N-2}) \, \mathrm{d}\varphi_{N-1} \cdots \mathrm{d}\varphi_2 \, \mathrm{d}\varphi_1 \mathrm{d}r$$

We can write the iterated integral above as a product of integrals

$$V_N(R) = \left(\int_0^R r^{N-1} \mathrm{d}r \right) \left(\int_0^{2\pi} \mathrm{d}\varphi_{N-1} \right) \left(\int_0^\pi \sin^{N-2}(\varphi_1) \mathrm{d}\varphi_1 \right)$$
$$\cdots \left(\int_0^\pi \sin(\varphi_{N-2}) \mathrm{d}\varphi_{N-2} \right)$$

Evaluating the first integral,

$$V_N(R) = \frac{R^N}{N} \left(\int_0^{2\pi} \mathrm{d}\varphi_{N-1} \right) \left(\int_0^\pi \sin^{N-2}(\varphi_1) \mathrm{d}\varphi_1 \right)$$
$$\cdots \left(\int_0^\pi \sin(\varphi_{N-2}) \mathrm{d}\varphi_{N-2} \right)$$

---

[1] Blumenson, L. E. (1960). *A Derivation of n-Dimensional Spherical Coordinates*. The American Mathematical Monthly. 67 (1): 63–66. doi:10.2307/2308932. JSTOR 2308932.

17.1. VOLUME OF A HYPERSPHERE OF DIMENSION N

Notice that this is very close to the trigonometric integral form of beta function,

$$B(x,y) = 2\int_0^{\frac{\pi}{2}} \sin^{2x-1}(t)\cos^{2y-1}(t)dt$$

See (11.3) for derivation. However, we need to find a way to change the interval of our integrals to $\left[0, \frac{\pi}{2}\right]$. The first integral's change in domain is trivial to compute. But, for the integrals with a sin term, we need to notice that $\sin^{N-1}(x)$ is symmetric around $\frac{\pi}{2}$ in the interval $[0, \pi]$.

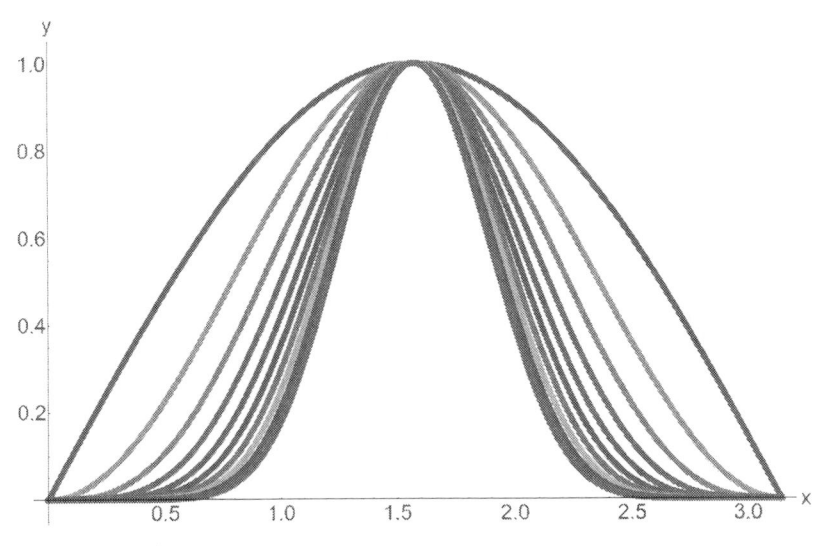

Figure 17.1: Graph of $y = \sin^N x$ for different values of N on $x \in [0, \pi]$

Therefore,

$$V_N(R) = \frac{R^N}{N}\left(2\int_0^{\frac{\pi}{2}}\sin^{N-2}(\varphi_1)\mathrm{d}\varphi_1\right)\cdots\left(2\int_0^{\frac{\pi}{2}}\sin(\varphi_{N-2})\mathrm{d}\varphi_{N-2}\right)$$
$$\cdot\left(4\int_0^{\frac{\pi}{2}}\mathrm{d}\varphi_{N-1}\right)$$

Now, we have a form that is equivalent to (11.3)! We can then write $V_N(R)$ as

$$V_N(R) = \frac{R^N}{N}\cdot\mathrm{B}\left(\frac{N-1}{2},\frac{1}{2}\right)\cdot\mathrm{B}\left(\frac{N-2}{2},\frac{1}{2}\right)\cdots\mathrm{B}\left(1,\frac{1}{2}\right)\cdot 2\mathrm{B}\left(\frac{1}{2},\frac{1}{2}\right)$$

$$= \frac{R^N}{N}\cdot\frac{\Gamma\left(\frac{N-1}{2}\right)\Gamma\left(\frac{1}{2}\right)}{\Gamma\left(\frac{N}{2}\right)}\cdot\frac{\Gamma\left(\frac{N-2}{2}\right)\Gamma\left(\frac{1}{2}\right)}{\Gamma\left(\frac{N-1}{2}\right)}\frac{\Gamma\left(\frac{N-3}{2}\right)\Gamma\left(\frac{1}{2}\right)}{\Gamma\left(\frac{N-2}{2}\right)}$$
$$\cdots\frac{\Gamma(1)\Gamma\left(\frac{1}{2}\right)}{\Gamma\left(\frac{3}{2}\right)}\cdot 2\frac{\Gamma\left(\frac{1}{2}\right)\Gamma\left(\frac{1}{2}\right)}{\Gamma(1)}$$

Notice that the finite product telescopes. Thus, using $n\Gamma(n) = \Gamma(n+1)$, we have

$$V_N(R) = \frac{2\pi^{\frac{N}{2}}R^N}{N\Gamma\left(\frac{N}{2}\right)}$$

$$\therefore V_N(R) = \frac{\pi^{\frac{N}{2}}R^N}{\Gamma\left(\frac{N}{2}+1\right)} \qquad (17.1)$$

17.1. VOLUME OF A HYPERSPHERE OF DIMENSION N 411

17.1.3 Discussion

Using (17.1), the unit ball in N dimensions has volume

$$V_N(1) = \frac{\pi^{\frac{N}{2}}}{\Gamma\left(\frac{N}{2} + 1\right)}$$

An interesting question might be:

> When does the volume of a unit N−ball reach its maximum?

We can plot $V_N(1)$ to get an idea:

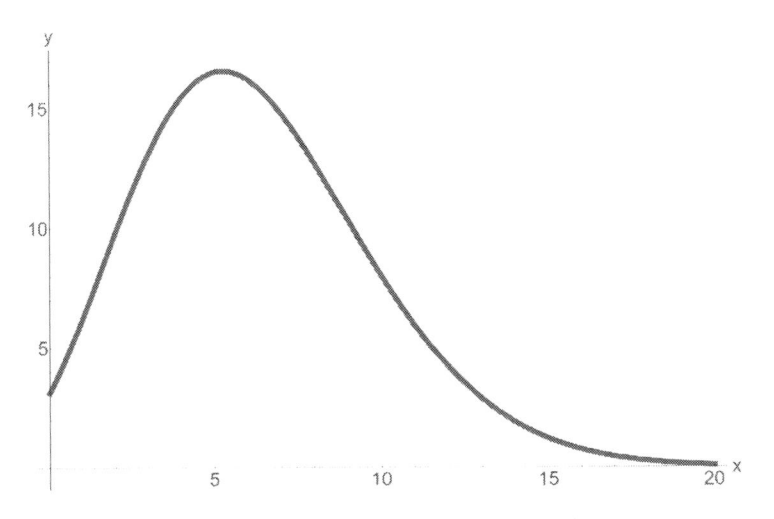

Figure 17.2: Graph of $y = V_x(1)$

The volume demonstrably peaks at $N = 5$. But, what is so special about 5? Nothing particular. If the radius were to be changed, the peak in (17.2) would shift. Consider, for example, an N−ball of radius 2 (figure (17.3)).

412   CHAPTER 17. MISCELLANEOUS

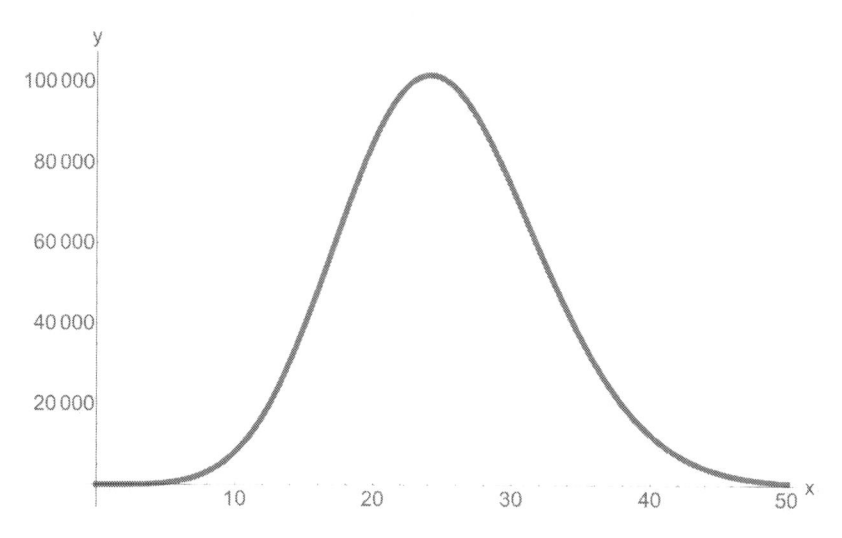

Figure 17.3: For the case of $R = 2$, the volume peaks around $N = 25$

Which demonstrably peaks much later. This trend of larger peaks should be intuitive from the R^N term in the numerator.

> **Challenge Problem**
>
> Let $n \in \mathbb{N}$ be the value at which $V_N(R)$ has a maximum. Prove that:
>
> $$\lfloor 2\pi R^2 - 2e^{-\gamma} \rfloor \leq n \leq \lfloor 2\pi R^2 - 1 \rfloor$$
>
> Hint: Use the result:[a,b]
>
> $$\log\left(x - \frac{1}{2}\right) < \psi(x) < \ln\left(x + e^{-\gamma} - 1\right)$$
>
> ---
> [a] F. Qi and B.-N. Guo. *Sharp inequalities for the psi function and harmonic numbers.* arXiv:0902.2524.
>
> [b] N. Elezovic, C. Giordano and J. Pecaric. *The best bounds in Gautschi's inequality.* Math. Inequal. Appl. 3 (2000). 239–252.

However, one feature for all radii holds true, the volume of the

17.1. VOLUME OF A HYPERSPHERE OF DIMENSION N

hypersphere *always* tends to 0 as $N \to \infty$. This is easy to show by taking the limit of $V_N(R)$:

$$\lim_{N \to \infty} V_N(R) = \lim_{N \to \infty} \frac{\pi^{\frac{N}{2}} R^N}{\Gamma\left(\frac{N}{2} + 1\right)}$$

Which can be easily shown to equal 0 by using Stirling's formula. But, what is an intuitive explanation for this? From the definition of an N−ball, we know that the region defined by

$$x_1^2 + x_2^2 + \cdots x_N^2 \leq R^2$$

Gives the volume of a N−ball of radius R. For the sake of simplicity, let us consider $R = 1$. This argument can, however, easily be extended to any $R > 0$. As N grows, most x_n must be close to 0. Consider the line given by:

$$x_1 = x_2 = \cdots = x_N$$

This line intersects our N−ball at $\pm \left(\frac{1}{\sqrt{N}}, \cdots, \frac{1}{\sqrt{N}}\right)$. As $N \to \infty$, this bounding region becomes smaller and smaller.

17.1.4 Applications

This brings us to the importance of this rather abstract problem. In fact, the result we obtained about the limiting behavior of $V_N(R)$ is a part of the larger **curse of dimensionality**:

Definition

The curse of dimensionality refers to various phenomena that arise when analyzing data in high-dimensional spaces, that do not occur in low-dimensinal space such as the 3−dimensional space we experience everyday. This expression was coined by the American mathematician Richard E. Bellman. Bellman used this expression to characterize some phenomena in *dynamic programming*, a method developed by Bellman[a,b].

[a]Richard Ernest Bellman (1961). *Adaptive control processes: a guided tour*. Princeton University Press.
[b]Richard Ernest Bellman (2003). *Dynamic Programming*. Courier Dover Publications. ISBN 978-0-486-42809-3.

This phenomenon arises often in statistics and machine learning. Interestingly, with a fixed amount of training data, the predictive power of a machine learning classifier or regressor first increases but then decreases. This phenomenon is named the *peaking phenomenon* and is similar to the trend in volume of a $N-$ball (See figure 17.2 and 17.3)[2]. Indeed, it is because the notion of $N-$balls is vital to data analysis in higher dimensions. Our analysis of $N-$balls is especially relevant to $k-$nearest neighbor classification[3].

17.1.5 Mathematical Connections

The notion of the volume of a $N-$ball appears in many fields of mathematics, especially in measure theory. For example, for any Borel set S, the following equality holds:

[2]Koutroumbas, Sergios Theodoridis, Konstantinos (2008). *Pattern Recognition - 4th Edition*. Burlington.
[3]Radovanović, Miloš; Nanopoulos, Alexandros; Ivanović, Mirjana (2010). *Hubs in space: Popular nearest neighbors in high-dimensional data*. Journal of Machine Learning Research. 11: 2487–2531.

17.1. VOLUME OF A HYPERSPHERE OF DIMENSION N

$$\lambda_N(S) = 2^{-N} V_N(1) H^N(S)$$

Where

- λ_N denotes the N-dimensional Lebesgue measure
- $H^N(S)$ denotes the N-dimensional Hausdorff measure

The reader should be familiar with the Lebesgue measure, which was discussed briefly earlier. The Hausdorff measure is fundamental in geometric measure theory, and appears in harmonic analysis[4,5].

---

[4]Morgan, F. (2016). *Geometric measure theory: a beginners guide*. Amsterdam: Academic Press.

[5]Toro, T. (2019). *Geometric Measure Theory–Recent Applications*. Notices of the American Mathematical Society, 66(04), 1. doi: 10.1090/noti1853

Part V

Appendices

Appendix A

It is standard knowledge that the imaginary unit, i, is equal to $\sqrt{-1}$. The term "imaginary" was coined by the French polymath René Descartes in an effort to devalue imaginary numbers' validity[6]. In the third book of his *La Géométrie*, Descartes states:

> Moreover, the true roots as well as the false [roots] are not always real; but sometimes only imaginary [quantities]; that is to say, one can always imagine as many of them in each equation as I said; but there is sometimes no quantity that corresponds to what one imagines, just as although one can imagine three of them in this [equation], $x^3 - 6x^2 + 13x - 10 = 0$, only one of them however is real, which is 2, and regarding the other two, although one increase, or decrease, or multiply them in the manner that I just explained, one would not be able to make them other than imaginary [quantities].

However, after the pioneering work of Euler and Gauss, both acceptance and interest in imaginary numbers rose.

Interest in e and the imaginary unit first arose from Bernoulli's

---

[6]Giaquinta, Mariano; Modica, Giuseppe (2004). *Mathematical Analysis: Approximation and Discrete Processes.* Springer Science Business Media. p. 121. ISBN 978-0-8176-4337-9. Extract of page 121

observation that:[7]

$$\frac{1}{1+x^2} = \frac{1}{2}\left(\frac{1}{1+ix} + \frac{1}{1-ix}\right)$$

Which, when combined with the standard integral

$$\int \frac{dx}{1+\alpha x} = \frac{\ln(1+\alpha x)}{\alpha} + C$$

Can give us an idea of what the complex natural logarithm is. Another notable formula is due to the English mathematician Roger Cotes:[8]

$$\ln(\cos x + i \sin x) = ix$$

Which was discovered in 1714. In 1748, Euler published his famous formula which states that

$$e^{ix} = \cos x + i \sin x$$

Proof. We will present a standard proof due to Euler. Recall the Taylor series for $\sin x$ and $\cos x$:

$$\sin x = \sum_{n=0}^{\infty} \frac{(-1)^n x^{2n+1}}{(2n+1)!} = x - \frac{x^3}{3!} + \frac{x^5}{5!} + \cdots$$

$$\cos x = \sum_{n=0}^{\infty} \frac{(-1)^n x^{2n}}{(2n)!} = 1 - \frac{x^2}{2!} + \frac{x^4}{4!} + \cdots$$

Also, recall the Taylor series of the exponential function

---

[7]Bernoulli, Johann (1702). *Solution d'un problème concernant le calcul intégral, avec quelques abrégés par rapport à ce calcul* [Solution of a problem in integral calculus with some notes relating to this calculation]. Mémoires de l'Académie Royale des Sciences de Paris. 1702: 197–289.

[8]Grattan-Guinness, I. (2000). *The rainbow of mathematics: a history of the mathematical sciences.* New york: W.W. Norton Company.

$$e^x = \sum_{n=0}^{\infty} \frac{x^n}{n!} = 1 + x + \frac{x^2}{2!} + \cdots$$

Substituting $x \to ix$ in the series expansion of e^x gives

$$e^{ix} = \sum_{n=0}^{\infty} \frac{(ix)^n}{n!}$$

$$= 1 + ix + \frac{i^2 x^2}{2!} + \frac{i^3 x^3}{3!} + \frac{i^4 x^4}{4!} + \cdots$$

$$= 1 + ix - \frac{x^2}{2!} - \frac{ix^3}{3!} + \frac{x^4}{4!} + \cdots$$

Notice that we can separate even and odd powers of x,

$$e^{ix} = \left(1 - \frac{x^2}{2!} + \frac{x^4}{4!} + \cdots\right) + i\left(x - \frac{x^3}{3!} + \frac{x^5}{5!} + \cdots\right)$$

The first sum is $\cos x$ and the second sum is $\sin x$. Therefore,

$$e^{ix} = \cos x + i \sin x \qquad (17.2)$$

□

Substituting $x \to -x$ in (17.2),

$$e^{-ix} = \cos x - i \sin x \qquad (17.3)$$

Since $\cos(\cdot)$ and $\sin(\cdot)$ are even and odd functions, respectively. Adding (17.2) and (17.3) gives

$$\cos x = \frac{e^{ix} + e^{-ix}}{2}$$

While subtracting (17.2) and (17.3) gives

$$\sin x = \frac{e^{ix} - e^{-ix}}{2i}$$

We can also derive the beautiful result

$$e^{i\pi} + 1 = 0$$

Using (17.2), which is famously known as Euler's identity. The physicist Richard Feynman called this equation "our jewel" and "the most remarkable formula in mathematics"[9]. The author agrees wholeheartedly with Feynman's assessment! It is perhaps useful to introduce another approach to deriving Euler's formula.

Proof. Let

$$f(x) = \cos x + i \sin x$$

We will begin by differentiating $f(x)$:

$$\frac{df}{dx} = -\sin x + i \cos x$$

$$= i f(x)$$

We now have the differential equation

---
[9]Feynman, Richard P. (1977). *The Feynman Lectures on Physics, vol. I*. Addison-Wesley. p. 22-10. ISBN 0-201-02010-6.

$$\frac{df}{dx} = if(x)$$

$$\frac{df}{f(x)} = i dx$$

Integrating,
$$\int \frac{df}{f(x)} = \int i \, dx$$

$$\ln f(x) = ix + C$$
$$\implies f(x) = e^{ix+C} = C_1 e^{ix}$$

Now, $f(0) = 1$, so $C_1 = 1$. Therefore,

$$e^{ix} = \cos x + i \sin x$$

$\square$

We now move our discussion to the complex logarithm.

> **Definition**
>
> The complex logarithm is defined similarly to the real logarithm[a]. Let
> $$e^x = z$$
> Then
> $$\ln z = x$$
>
> ---
> [a]Sarason, Donald (2007). *Complex Function Theory (2nd ed.)*. American Mathematical Society.

Proposition. For any complex number, there are infinitely many complex logarithms.

Proof. Let $z = re^{i\theta}$ for $r, \theta \in \mathbb{R}$ and $r > 0$. Then one value of the complex logarithm is:

$$\ln z = \ln r + i\theta$$

However, note that z can also be represented as $z = re^{i\theta + 2ik\pi}$ for any $k \in \mathbb{N}$ since

$$e^{2ik\pi} = \cos(2\pi k) + i\sin(2\pi k) = 1$$

Taking the complex logarithm of both sides of $z = re^{i\theta + 2ik\pi}$ gives,

$$\ln z = \ln r + (\theta + 2k\pi)i$$

Since we can choose any $k \in \mathbb{N}$, there are infinitely many values of the complex logarithm. $\square$

The complex logarithm is defined as a *multi-valued* function. Hence, a **branch** is required to define the complex logarithm as a function. In most cases, we will deal with the *principal value* of the complex logarithm. For each complex number $z \in \mathbb{C} \setminus 0$, the principal value of $\ln z$ is the logarithm whose imaginary part lies in the interval $(-\pi, \pi]$. We can then write the principal value of the complex logarithm as

$$\ln z = \ln r + \left\{\frac{\theta}{2\pi}\right\} \tag{17.4}$$

For some $z = re^{i\theta}$. Here $\{\cdot\}$ denotes the fractional part. The interested reader should refer to chapter 8 for a survey of problems involving fractional parts.

Appendix B

Any introductory two-semester calculus course sequence is bound to investigate Taylor series. However, this appendix shall serve as a quick review.

The Taylor series of some function, $f(x)$, is its representation as an infinite sum of terms. These terms are calculated from the values of the derivatives of $f(x)$ at a single point. We can present a more formal definition below:

Definition

Let $f(x)$ be an infinitely differentiable complex-valued function at $x = \alpha$. Its Taylor series is then given by the infinite power series:

$$f(x) = f(\alpha) + \frac{f'(\alpha)}{1!}(x-\alpha) + \frac{f''(\alpha)}{2!}(x-\alpha)^2 + \frac{f'''(\alpha)}{3!}(x-\alpha)^3 + \cdots$$

Or, using summation notation,

$$\sum_{n=0}^{\infty} \frac{f^{(n)}(\alpha)}{n!}(x-\alpha)^n$$

Where $f^{(n)}$ denotes the nth derivative of f. Some Taylor series include:

Exponential Function

$$e^x = \sum_{n=0}^{\infty} \frac{x^n}{n!}$$

Which converges for all $x \in \mathbb{C}$

Logarithmic Function

$$\ln(1-x) = -\sum_{n=1}^{\infty} \frac{x^n}{n} = -x - \frac{x^2}{2} - \frac{x^3}{3} - \cdots,$$

$$\ln(1+x) = \sum_{n=1}^{\infty} \frac{(-1)^{n-1} x^n}{n} = x - \frac{x^2}{2} + \frac{x^3}{3} - \cdots.$$

Which always converge for $|x| < 1$. It is worth noting that the series representations of $\ln(1-x)$ and $\ln(1+x)$ converge at $x = -1$ and $x = 1$, respectively.

Trigonometric Functions

$$\sin x = \sum_{n=0}^{\infty} \frac{(-1)^n}{(2n+1)!} x^{2n+1} = x - \frac{x^3}{3!} + \frac{x^5}{5!} - \cdots$$

$$\cos x = \sum_{n=0}^{\infty} \frac{(-1)^n}{(2n)!} x^{2n} = 1 - \frac{x^2}{2!} + \frac{x^4}{4!} - \cdots$$

$$\tan x = \sum_{n=1}^{\infty} \frac{B_{2n}(-4)^n (1-4^n)}{(2n)!} x^{2n-1} = x + \frac{x^3}{3} + \frac{2x^5}{15} + \cdots$$

$$\arcsin x = \sum_{n=0}^{\infty} \frac{(2n)!}{4^n (n!)^2 (2n+1)} x^{2n+1} \qquad = x + \frac{x^3}{6} + \frac{3x^5}{40} + \cdots$$

$$\arccos x = \frac{\pi}{2} - \arcsin x$$

$$= \frac{\pi}{2} - \sum_{n=0}^{\infty} \frac{(2n)!}{4^n (n!)^2 (2n+1)} x^{2n+1} \qquad = \frac{\pi}{2} - x - \frac{x^3}{6} - \frac{3x^5}{40} - \cdots$$

$$\arctan x = \sum_{n=0}^{\infty} \frac{(-1)^n}{2n+1} x^{2n+1} \qquad = x - \frac{x^3}{3} + \frac{x^5}{5} - \cdots$$

Where B_n denotes the nth Bernoulli number, see (4.4) for more.

Hyperbolic Trigonometric Functions

$$\sinh x = \sum_{n=0}^{\infty} \frac{x^{2n+1}}{(2n+1)!} \qquad = x + \frac{x^3}{3!} + \frac{x^5}{5!} + \cdots$$

$$\cosh x = \sum_{n=0}^{\infty} \frac{x^{2n}}{(2n)!} \qquad = 1 + \frac{x^2}{2!} + \frac{x^4}{4!} + \cdots$$

$$\tanh x = \sum_{n=1}^{\infty} \frac{B_{2n} 4^n (4^n - 1)}{(2n)!} x^{2n-1} \qquad = x - \frac{x^3}{3} + \frac{2x^5}{15} - \frac{17x^7}{315} + \cdots$$

$$\sinh^{-1} x = \sum_{n=0}^{\infty} \frac{(-1)^n (2n)!}{4^n (n!)^2 (2n+1)} x^{2n+1} \qquad = x - \frac{x^3}{6} + \frac{3x^5}{40} + \cdots$$

$$\tanh^{-1} x = \sum_{n=0}^{\infty} \frac{x^{2n+1}}{2n+1} \qquad = x + \frac{x^3}{3} + \frac{x^5}{5} + \cdots$$

Again, B_n denotes the nth Bernoulli number.

Binomial Series

$$(1+x)^\alpha = \sum_{n=0}^{\infty} \binom{\alpha}{n} x^n$$

Where $\binom{\alpha}{n}$ denotes the generalized binomial coefficient,

$$\binom{\alpha}{n} = \prod_{k=1}^{n} \frac{\alpha - k + 1}{k} = \frac{\alpha(\alpha - 1) \cdots (\alpha - n + 1)}{n!}$$

Notice that when $\alpha \in \mathbb{Z}^-$, this series is equivalent to the series for

$$\frac{1}{1+x}, \frac{1}{(1+x)^2}, \frac{1}{(1+x)^3}, \cdots$$

To derive many of the above series, one needs to employ the technique of interchanging summation and integration. A nice challenge would be trying to derive all the series listed above. Give it a try!

Acknowledgements

First and foremost, I would like to thank everyone in the Daily Math community. My followers and supporters on the Instagram page @daily_math_ have been instrumental in my motivation to complete this book. At the time of publishing this book, the Daily Math community is over 40,000 members strong and is growing at an amazing rate. From this amazing community, a group of individuals with varying backgrounds were selected to give feedback on the book before it is published. I am very grateful for their feedback, as they have truly helped shape this book. These individuals, in no particular order, are

- Poon Zong Wei Julian
- Anton Steinfadt
- Duc Van Khanh Tran
- Solden Stoll
- Kevin Tong
- Jack Moffatt
- Vasilij Shikunov
- Florian Babisch
- Keyvon Rashidi

- Others who did not want their name to be listed

I would like to thank everyone who assisted in the publishing of this book. I am indebted to my graphic designer, Ayan Rasulova, for her great efforts in designing the book cover and for Christian Cortez for his help in the design of the book cover.

Last, but certainly not least, I would like to express my gratitude to all the amazing people in my life that motivated me to start and finish this book. I truly would not have completed this journey without my parents, teachers, and friends always on my side. I would like to express my sincere gratitude to my calculus teacher, Mary Rubin, for her role in developing my mathematical interests and to my chemistry teacher, Maliha Malik, for her support throughout this journey.

A special thank you is owed to Victor Del Carpio, who inspired me to start this book and Hrishik Rangaraju, who always motivated me to complete the book. Their role as friends and peers throughout this journey helped me in keeping my goals in sight through the many ups and downs.

I am extremely grateful to my parents for their love, prayers, and sacrifices to give me the best education possible. Their hard work and dedication to my success has shaped my life, and with their help I navigated the many obstacles in writing this book.

Answers

Chapter 1

4. 1

5. 1

6. -1

7. $-\dfrac{2}{e}$

8. $\dfrac{1}{2019}$

9. 1

10. $\dfrac{64}{27}$

11. $\dfrac{\pi}{2}$

Chapter 2

1. $\dfrac{\pi\sqrt{3} - 3\ln 2}{9}$

2. $\dfrac{\pi \ln 2}{8}$

3. π

4. $\dfrac{\sqrt{3} - \sqrt{2}}{2}$

5. $4\sqrt{2} - 4$

6. $\dfrac{\pi}{4}$

7. $\dfrac{\pi \ln 2}{3\sqrt{3}}$

8. $\left(3\sqrt{2} - 2\right)\pi$

9. The integrand is always positive on $[0, 1]$, so the integral is positive. Moreover, the integral equals $\frac{22}{7} - \pi$, which shows that $\frac{22}{7} > \pi$.

10. $\dfrac{e^{\frac{1}{4}\pi}}{2}$

Chapter 3

1. $\ln 3$

2. $\dfrac{\pi \ln 2}{8}$

3. $\pi \ln\left(\dfrac{\sqrt{\alpha}+\sqrt{\beta}}{2}\right)$

4. $\dfrac{\pi}{2}$

5. 2π

7. $\dfrac{\pi \ln 2}{2}$

8. $\dfrac{\pi}{2e}$

9. $\dfrac{\pi}{4\sqrt{\alpha\beta}}\left(\dfrac{1}{\alpha}+\dfrac{1}{\beta}\right)$

Chapter 4

2. $\dfrac{11}{96}$

3. $\dfrac{2019}{1010}$

4. $(k+1)(k+1)! - 1$

6. $\dfrac{3\sin 1}{4}$

7. $\ln 2 - \ln 2020 - \ln(2021!)$

Chapter 6

1. $\dfrac{1}{2e}$

2. $\dfrac{179\pi}{360}$

3. $\dfrac{\pi}{2}$

4. $\dfrac{(\ln 2)^2}{2} - \dfrac{\zeta(2)}{2}$

5. $\dfrac{2k+2}{k+2}$

Chapter 7

1. $2 - \dfrac{\pi^2}{6}$

2. $\dfrac{\pi}{2^{2n+1}}\binom{2n}{n}$

3. $\dfrac{\pi^2}{2}$

4. $-G$

5. $\dfrac{7}{4}\zeta(3)$

6. $\dfrac{7}{32}\zeta(3)$

7. $\ln 2$

8. $\pi \left[\ln\left(\dfrac{\sqrt{2}+2}{4}\right) + \dfrac{\pi}{4} + 1 - \sqrt{2} \right]$

Chapter 8

1. $-\dfrac{2n^3}{3} + \sum_{k=1}^{n^2} \sqrt{k}$

2. $\dfrac{1}{2}$

3. $\dfrac{1}{2020^2} - \dfrac{\zeta(2021)}{(2020)(2021)}$

4. $\dfrac{\zeta(2)-1}{2}$

5. π

Chapter 9

2. $\dfrac{\ln 2\pi}{4} + \dfrac{1}{8}$

4. $5!\,\zeta(4)$

5. $\dfrac{144}{625}$

6. $\dfrac{\ln^2 2\pi}{3} + \dfrac{\pi^2}{48} + \dfrac{\gamma \ln\sqrt{2\pi}}{3} + \dfrac{\gamma^2}{12} + \dfrac{\zeta''(2)}{2\pi^2} - \dfrac{\ln(2\pi)\zeta'(2)}{\pi^2} - \dfrac{\gamma\zeta'(2)}{\pi^2}$

Chapter 10

1. $-\gamma$

2. $3 - 2\gamma$

4. $2\zeta(3)$

6. $\dfrac{3\pi^2}{2}G - \dfrac{1}{128}\left(\psi^{(3)}(1/4) - \psi^{(3)}(3/4)\right)$

Chapter 11

2. $\sqrt{\dfrac{2}{\pi}}\left(\Gamma\left(\dfrac{3}{4}\right)\right)^2$

3. $\dfrac{\pi^3}{16}$

6. $\dfrac{\sqrt{\pi}\,\Gamma(5/4)}{2\Gamma(3/4)}$

7. $\dfrac{1}{4}\left(\Gamma(1/4)\right)^4$

Chapter 12

1. 1

3. $\dfrac{\pi^2}{6}$

4. $\dfrac{\pi^2}{6} - \zeta(3)$

5. $1 - \gamma$

6. $-\dfrac{5}{8}\zeta(3)$

7. $\dfrac{7}{16}\zeta(3) - \dfrac{\pi^2 \ln 2}{8}$

Integral Table

The integration constant has been omitted here.

Rational Functions

$$\int x^n \, \mathrm{d}x = \frac{1}{n+1} x^{n+1}, \quad n \neq -1$$

$$\int \frac{1}{x} \, \mathrm{d}x = \ln|x|$$

Exponential Functions

$$\int e^x \, \mathrm{d}x = e^x$$

$$\int a^x \, \mathrm{d}x = \frac{1}{\ln a} a^x$$

Logarithmic Function

$$\int \ln x \, dx = x \ln x - x$$

Trigonometric Functions

$$\int \sin x \, dx = -\cos x$$

$$\int \cos x \, dx = \sin x$$

$$\int \tan x \, dx = \ln|\sec x|$$

$$\int \sec x \, dx = \ln|\sec x + \tan x|$$

$$\int \sec^2 x \, dx = \tan x$$

$$\int \sec x \tan x \, dx = \sec x$$

Miscellaneous

$$\int \frac{a}{a^2 + x^2} dx = \tan^{-1} \frac{x}{a}$$

$$\int \frac{a}{a^2 - x^2} dx = \frac{1}{2} \ln \left| \frac{x+a}{x-a} \right|$$

$$\int \frac{1}{\sqrt{a^2 - x^2}} dx = \sin^{-1} \frac{x}{a}$$

$$\int \frac{a}{x\sqrt{x^2 - a^2}} dx = \sec^{-1} \frac{x}{a}$$

$$\int \frac{1}{\sqrt{x^2 - a^2}} dx = \cosh^{-1} \frac{x}{a}$$
$$= \ln\left(x + \sqrt{x^2 - a^2}\right)$$

$$\int \frac{1}{\sqrt{x^2 + a^2}} dx = \sinh^{-1} \frac{x}{a}$$
$$= \ln\left(x + \sqrt{x^2 + a^2}\right)$$

Trigonometric Identities

Unit Circle Identities

$$\sin^2(\theta) + \cos^2(\theta) = 1$$

Dividing $\sin^2(\theta) + \cos^2(\theta) = 1$ by $\cos^2(\theta)$ or $\sin^2(\theta)$ gives

$$\tan^2(\theta) + 1 = \sec^2(\theta)$$
$$1 + \cot^2(\theta) = \csc^2(\theta)$$

Addition Formulas

$$\sin(\alpha + \beta) = \sin(\alpha)\cos(\beta) + \cos(\alpha)\sin(\beta)$$
$$\cos(\alpha + \beta) = \cos(\alpha)\cos(\beta) - \sin(\alpha)\sin(\beta)$$

Double Angle Formulas

$$\sin(2\alpha) = 2\sin(\alpha)\cos(\alpha)$$
$$\cos(2\alpha) = \cos^2(\alpha) - \sin^2(\alpha)$$

The formula for $\cos(2\alpha)$ is often rewritten by replacing $\cos^2(\alpha)$ with $1 - \sin^2(\alpha)$ or replacing $\sin^2(\alpha)$ with $1 - \cos^2(\alpha)$ to get

$$\cos(2\alpha) = 1 - 2\sin^2(\alpha)$$
$$\cos(2\alpha) = 2\cos^2(\alpha) - 1$$

Solving for $\sin^2(\alpha)$ and $\cos^2(\alpha)$ gives power reduction formulas that are very useful for integration

$$\sin^2(\alpha) = \frac{1}{2}(1 - \cos(2\alpha))$$
$$\cos^2(\alpha) = \frac{1}{2}(1 + \cos(2\alpha))$$

An addition formula for tangent can be derived from the ones for sine and cosine.

$$\tan(\alpha + \beta) = \frac{\sin(\alpha)\cos(\beta) + \cos(\alpha)\sin(\beta)}{\cos(\alpha)\cos(\beta) - \sin(\alpha)\sin(\beta)}$$

Now dividing by $\dfrac{\cos(\alpha)\cos(\beta)}{\cos(\alpha)\cos(\beta)}$ gives

$$\tan(\alpha + \beta) = \frac{\tan(\alpha) + \tan(\beta)}{1 - \tan(\alpha)\tan(\beta)}$$

$$\tan(2\alpha) = \frac{2\tan(\alpha)}{1 - \tan^2(\alpha)}$$

Triple Angle Formulas

$$\sin(3\theta) = 3\sin\theta - 4\sin^3\theta = 4\sin\theta \sin\left(\frac{\pi}{3} - \theta\right)\sin\left(\frac{\pi}{3} + \theta\right)$$

$$\cos(3\theta) = 4\cos^3\theta - 3\cos\theta = 4\cos\theta \cos\left(\frac{\pi}{3} - \theta\right)\cos\left(\frac{\pi}{3} + \theta\right)$$

$$\tan(3\theta) = \frac{3\tan\theta - \tan^3\theta}{1 - 3\tan^2\theta} = \tan\theta \tan\left(\frac{\pi}{3} - \theta\right)\tan\left(\frac{\pi}{3} + \theta\right)$$

Half Angle Formulas

$$\sin^2\left(\frac{\theta}{2}\right) = \frac{1 - \cos\theta}{2}$$

$$\cos^2\left(\frac{\theta}{2}\right) = \frac{1 + \cos\theta}{2}$$

$$\tan^2\left(\frac{\theta}{2}\right) = \frac{1 - \cos\theta}{\sin\theta}$$

Try deriving other trigonometric identities using these ones as a basis!

Alphabetical Index

A
Apery's constant 314
Augustin-Louis Cauchy .. 30

B
Basel problem 268, 315
Bernhard Riemann . 60, 314
Bernoulli numbers . 141, 427
Binomial theorem .. 45, 210, 427
Boltzmann's constant .. 380, 388, 393, 398
Brook Taylor 82

C
Carl Freidrich Gauss ... 136, 264, 280, 419
Catalan's constant 222, 275
Chain rule .. 50, 55, 57, 360
Crystal structure . 366, 370, 379
Curse of dimensionality 413

D
Dynamic programming . 414

E
Erwin Madelung 367
Euler's formula 200, 420
Euler's reflection formula
    71, 89, 267, 274, 305, 310

F
Fibonacci sequence . 44, 154
Fractional derivative 51
Fundamental theorem of arithmetic 320, 321

G
Gaussian integral .. 116, 271
Generating function ... 141, 206, 321
Geometric measure theory 415
Glaisher-Kinkelin constant 253, 330
Gottfried Wilhelm Leibniz 51, 60
Guillaume de L'Hopital . 36

H
Harmonic number 170,

447

204–206, 228
Hausdorff measure 415
Henri Lebesgue 63

I
Integration by parts 81

J
James Clerk Maxwell .. 386
Johann Bernoulli ... 36, 419
Josiah Willard Gibbs ... 386

K
Karl Weierstrass ... 30, 265, 269

L
L'Hopital's rule 47, 112, 145, 334, 381
Lagrange multiplier 349, 401
Lagrangian 348, 356
Lebesgue dominated convergence theorem ... 187, 188
Lebesgue integral 63, 64
Lebesgue measure . 187, 415
Legendre duplication formula 310
Leonhard Euler 71, 241, 268, 314, 315, 319, 420, 422
Lord Rayleigh 393

Ludwig Boltzmann 386

M
Max Planck 394

P
Pafnuty Chebyshev 319
Pauli exclusion principle 390, 397
Polar coordinates .. 199, 300

R
Raabe's integral 273
René Descartes 419
Richard Feynman . 102, 422
Riemann hypothesis 314
Riemann series theorem 171, 378
Riemann sums .. 60, 62, 63, 177
Roger Cotes 420

S
Special relativity 364
Spherical coordinates .. 312, 406, 408
Stefan–Boltzmann law . 397
Stirling's formula .. 43, 250, 400, 413

W
Weierstrass factorization theorem 269

Printed in Poland
by Amazon Fulfillment
Poland Sp. z o.o., Wrocław